WHO NEEDS CREDIT?

Who Needs Credit?
POVERTY AND FINANCE IN BANGLADESH

EDITED BY
Geoffrey D Wood & Iffath A Sharif

ZED BOOKS
London & New York

Who Needs Credit? Poverty and Finance in Bangladesh was first published by Zed Books Ltd, 7 Cynthia Street, London N1 9JF, UK and Room 400, 175 Fifth Avenue, New York, NY 10010, USA in 1997.

Distributed in the USA exclusively by St Martin's Press, Inc., 175 Fifth Avenue, New York, NY 10010, USA.

Published in South Asia by The University Press Ltd, Red Crescent Building, 114 Motijheel C/A, GPO Box 2611, Dhaka 1000, Bangladesh.

© The University Press Ltd, Dakha

Cover design by Andrew Corbett

Printed and bound in the United Kingdom by Biddles Ltd, Guildford and King's Lynn.

The moral rights of the authors of this work have been asserted by them in accordance with the Copyright, Designs and Patents Act, 1988.

A catalogue record of this book is available from the British Library.

Library of Congress Cataloging-in-Publication Data has been applied for.

ISBN 1 85649 523 X Hb
ISBN 1 85649 524 8 Pb

Contents

Contributors	13
Acknowledgement	19
Acronyms	21
Glossary	25

INTRODUCTION
Geoffrey D Wood and Iffath Sharif

Context: Micro-Credit Summit	27
Workshop on Poverty and Finance in Bangladesh	33
Borrower Sustainability: Introducing the Papers	38

PART I: OVERVIEW

Chapter 1

POVERTY AND FINANCE IN BANGLADESH: A NEW POLICY AGENDA
Iffath Sharif

Introduction	61
The 'New World of Micro Finance Evangelism'	62
The Poverty Setting	64
Whose Graduation: Borrowers' or Institutions'?	70
Challenges Ahead	76
Conclusion	81

Chapter 2

POVERTY AND WELL-BEING: PROBLEMS FOR POVERTY REDUCTION IN ROLE OF CREDIT
Martin Greeley

Overview	83
Introduction	83

Chapter 3

FINANCE FOR THE POOR OR POOREST? FINANCIAL INNOVATION, POVERTY AND VULNERABILITY
David Hulme and Paul Mosley

Introduction	97
Poverty, Vulnerability and Deprivation	98
Finance for the Poor: Improving Incomes	101
Finance for the Poor: Reducing Vulnerability	109
Finance for the Poor: Influences on other Forms of Deprivation	118
Reconceptualising 'Finance for the Poor' and its Policy Implications	125
Conclusion	128

Chapter 4

THE POLITICAL ECONOMY OF MICRO CREDIT   131
Rehman Sobhan

PART II: CASE STUDIES & THEMES

Chapter 5

EXPERIENCES AND CHALLENGES IN CREDIT & POVERTY ALLEVIATION PROGRAMMES IN BANGLADESH: THE CASE OF PROSHIKA
Yuwa Hedrick-Wong, Bosse Kramsjo, Asgar Ali Sabri

Overview	145
Proshika Development Strategy and Experience	146
Proshika's Credit Service	148
Empirical Evidence of Proshika's Impacts	148
Criteria for Assessing the Level of Optimal Social Development Services	152
Case Studies	153
Overall Comments	163
Challenges Ahead	167

Chapter 6

BRAC'S POVERTY ALLEVIATION PROGRAMME: WHAT IT IS AND WHAT IT ACHIEVED
A Mushtaque R Chowdhury, M Aminul Alam

Abstract	171
Introduction	171
Social Mobilization for Poverty Alleviation: A Holistic Approach	172
The Process of Social Mobilization	173
BRAC's Theory of Development: A Learning Approach	175
The Size of BRAC and Financial Sustainability	177
Salient Features of BRAC Programmes	179
Links with the Public Sector and Other NGOs	184
How Successful have these Efforts been?	185
Discussion	190

Chapter 7

ASA 'SELF-RELIANT DEVELOPMENT MODEL'
A K Aminur Rashid

Introduction	195
ASA's Internal Organization	196
Financial Flows, Operational and Financial Self-Sufficiency	197
Beneficiary Self-Reliance	198
Sustainability Concepts	200

Chapter 8

CREDIT FOR POVERTY ALLEVIATION IN BANGLADESH: PERFORMANCE OF PUBLIC SECTOR BANKS
Mosharraf Hossain Khan

Introduction and Background	203
The Role of the Banks in Dispensing Credit for Poverty Alleviation	204
Programs Implemented Individually by the Banks	206
Issues and Recommendations	210
Concluding Remarks	213

Chapter 9

GRAMEEN BANK: A CASE STUDY
Syed M Hashemi, Lamiya Morshed

Introduction	217
Explanations of Effectiveness	218
Organization Structure	221
Performance	222
Economic Impact	223
Social Impact	224
Conclusion	224

PART III: PROBLEMS OF REACHING THE POOREST

Chapter 10

MICRO-CREDIT PROGRAMMES: WHO PARTICIPATES AND WHAT DOES IT MATTER?
Hassan Zaman

Introduction	231
Initial Endowment: Are BRAC Member's 'Homogenous'?	232
The Determinants of BRAC Membership: A Multivariate Analysis	234
The Depth of Participation	239
The Determinants of 'Membership Depth'	241
Concluding Discussion	244

Chapter 11

THOSE LEFT BEHIND: A NOTE ON TARGETING THE HARDCORE POOR
Syed M Hashemi

Introduction: Poverty in Rural Bangladesh	249
Grameen Bank and Micro-Credit: The Rationale	250
The Credit Delivery Model	251
Reaching the Poor	252

The BRAC IGVGD Program	255
Conclusion	256

PART IV: MICRO-CREDIT-LIMITATIONS OF SCALE

Chapter 12
THE RENEGOTIATION OF JOINT LIABILITY: NOTES FROM MADHUPUR
Imran Matin

The Pristine Story	262
The Fuzzy Story	264
The Ruptures and Renegotiation of Joint Liability	267
Those Who Get it Wrong	269

Chapter 13
POVERTY, PROFITABILITY OF MICRO ENTERPRISES AND THE ROLE OF CREDIT
Rushidan Islam Rahman

Introduction	271
Why the Poorest are not Covered by NGO Credit: Existing Hypotheses	274
Demand for Micro-Credit and Exclusion of the Poorest	275
The Supply Side Reconsidered	281
Concluding Remarks	284

Chapter 14
BREAKING OUT OF THE GHETTO: EMPLOYMENT GENERATION AND CREDIT FOR THE POOR
Geoffrey D Wood

Introduction	289
Equity and Equality	291
Who are the 'Entrepreneurial Poor'?	292
The Problem of 'Ghetto Credit'	294
Linking Credit and Employment	295

Significance of Macro Economic Performance and Policy 296
Institutional Options Linking Credit and Employment 298

PART V: MICRO-CREDIT: A RESTRICTED APPROACH TO FINANCIAL SERVICES

Chapter 15

SAVINGS: FLEXIBLE FINANCIAL SERVICES FOR THE POOR
Graham Wright, Mosharrof Hossain and Stuart Rutherford

Abstract	309
Introduction	310
Savings—Abroad	311
Savings—At Home	312
Description of BURO, Tangail's System	323
The Aims of the Study	325
Methods	325
Results and Discussion—Quantitative Survey (Summary Analysis Attached)	326
Results and Discussion—Qualitative Survey	331
Concluding Discussion	334
Summary Conclusions	336

Chapter 16

INFORMAL FINANCIAL SERVICES IN DHAKA'S SLUMS
Stuart Rutherford

Abstract	351
The Schemes	351
The 'Fund' *Samities*	362
The Wider Perspective	365
Managing the Capacity to Save	367
Informal Samities and the NGOs	368

CONCLUSION
Iffath Sharif and Geoffrey D Wood

Context	371

Current Achievements and Constraints 373
Thinking Differently about Poverty and Finance 375
Conclusion: Borrowers' Sustainability 378

References 381

Index 393

Contributors

A K M Aminur Rashid born in 1964 is a Master of Science from Dhaka University. As an Associate Coordinator of ASA (Association for Social Advancement) he is closely associated with the implementation of rural micro-finance program. He was deputed to HEKS-Cambodia program (A Swiss humanitarian NGO) as an expatriate in 1993. He discharged the responsibilities of 'preparatory formation of Credit and Savings Program in project area' in Cambodia. He was also hired by Save the Children/USA as an consultant of Group Guaranteed Lending and Savings (GGLS) program and worked in the Republic of Tajikistan from 1995 to 1996. He obtained a certificate in Training and Development from the Institute of Personnel and Development, UK, by completing a 'Training of Trainers' course arranged by Arhus Technical College, Denmark.

A Mushtaque R Chowdhury has been at BRAC since the mid 1970's and is Director of the Research and Evaluation Division at BRAC. He has researched and published extensively in the fields of health, education and credit. Dr. Chowdhury has a Masters degree from the London School of Economics, a Ph.D. from the London School of Hygiene and Tropical Medicine and worked as a post doctoral fellow at Harvard.

Asgar Ali Sabri completed his Honours and Masters in Social Welfare from the University of Dhaka. He Joined Proshika, a leading national NGO in Bangladesh as Junior Researcher in 1987. Later on, he completed Masters of Arts in Development Studies from the Institute of Social Studies (ISS), The Hague, The Netherlands and resumed his responsibility as the Co-ordinator of Impact Monitoring and Evaluation Cell (IMEC) of Proshika. He has co-authored the case of Proshika Livestock and Social Forestry Programmes in the book titled "*NGOs and State in South Asia*" edited by David J. Lewis and John Farrington, 1993. Presently, he is working as research fellow in the Policy Research Department (PRD) of IDPAA (Institute for Development Policy Analysis and Advocacy), Proshika.

Bosse Kramsjo, born in Sweden 1951, has been a teacher and staff member of Sando U-Centrum since 1986, Swedish Board for Education in International Development, Sando, Sweden. He studied International Development and Social Anthropology in the University of Gothenburg. He also completed 1 year combined theoretical and practical forestry training from the Swedish Agricultural University. Earlier he worked as an advisor to Proshika and did the liaison work on training and information between Swallows IRWP/RESP and SIDA, based in Dhaka. He has published and co-edited 3 books on international development issues including "Breaking the Chains-Collective Action for Social Justice Among the Rural Poor in Bangladesh".

David Hulme is Professor of Development Studies and Director of the Institute for Development Policy and Management at the University of Manchester. His main research interests are in rural development and poverty reduction strategies. He has researched and worked in Bangladesh, Sri Lanka, Kenya, Malawi and Papua New Guinea.

Geoffrey D Wood BA, MPhil, PhD, Reader in Development Studies and Director, Centre for Development Studies, School of Social Science, University of Bath. He has published 6 books (some co-authored) and numerous articles on agrarian change and rural development in the Indian sub-continent (mainly Bangladesh, but also India, Nepal, Sri Lanka and Pakistan), as well as working briefly in Thailand. His first research was on development administration issues in Zambia. He has worked in the Indian sub-continent for the last 25 years, spending proportions of each year there. In addition to primary research, he has advised many programmes for donors (specially SIDA, but more recently UK-ODA) as well as large NGOs and the Government of Bangladesh (supported by the Ford Foundation). He is currently researching on 'urban poor livelihood strategies in Dhaka slums'. He is also working on livelihoods and environmental issues in Bangladesh (flooding and water management), Ghana (environmental management systems in mining with colleagues and research students) and in Latin America (Venezuela, Peru and Colombia).

Hassan Zaman has been working as Senior Staff Economist at BRAC for the last three years and is also concurrently doing his doctorate in Economics under Professor Michael Lipton at the

University of Sussex. His thesis centres around the impact of BRAC's credit program on poverty alleviation and empowerment of the rural poor. He has both undergraduate and Masters degrees in Economics from the London School of Economics.

Iffath A Sharif is one of the Research Associates, working in the Policy Research Department of the Institute for Development Policy Analysis and Advocacy, Proshika. She is an economist by training and is currently working for a Masters in Public Affairs at the Woodrow Wilson School of Public and International Affairs at Princeton University. She has also served as a consultant to the United Nations Development Programme in Dhaka, working primarily on the UNDP strategy in the area of micro credit. Ms. Sharif started working on micro credit issues as a student and wrote her Bachelor's thesis on the role of micro credit, using experiences of the Grameen Bank. She is currently working on financial sector reform issues and their relevance to the sustainability of lending systems for the poor

Imran Matin born in 1970. Schooling in Faujdarhat Cadet College, Bangladesh. Attended Delhi University and Sussex University for an undergraduate economics and a Masters' in development economics degree respectively. At present studying for a Ph.D. in Economics in Sussex University. Research interest around the issues of microfinance delivery models and group dynamics.

Lamiya Morshed is Programme Officer at the Grameen Trust.

M Aminul Alam has been at BRAC since the mid 1970's. He has extensive experience in rural development and is currently Director of Field Operations of BRAC's largest programme, the Rural Development Programme.

Martin Greeley is an Economist, working on rural development with special emphasis on employment, income distribution and poverty reduction. He has undertaken long term research on the socioeconomic consequences of rural technological change in India, Bangladesh and Sri Lanka; other research interests include rural farm and nonfarm credit, incorporation of environmental considerations into poverty-focused development programmes, potential role of agricultural biotechnology for small farms in ldcs, currently researching economic growth and poverty reduction in rural Bangladesh. He is the Chairman of the Development Studies Graduate Research

Centre and, is also the Co-Director of the M.Phil. in Development Studies 1996-98.

Mosharraf Hossain Khan completed his Honours and Masters in Agricultural Science from Bangladesh Agricultural University, Mymensing, securing first class in both the exams; he started his career in rural banking as Junior Officer in Sonali Bank in 1974 and served in various places of the country in different capacities till 1994 when he assumed the charge of Deputy General Manager of Rural Credit Division of the Bank; he obtained Diploma in Banking from the Institute of Bankers, Bangladesh with distinction in Part-1 Exam.; a contributor of articles on rural banking and finance in different dailies, Mr. Khan worked as Credit Consultant/Expert in several donor assisted poverty focused Projects like RD-12 of CIDA, RD-5 of SIDA, DTW-2 Project of IDA and Rural Poor Cooperative Project of ADB; he visited India and the Philippines; he joined Palli Karma Sahayak Foundation (PKSF) as General Manager (Operations) in June 1996.

Mosharrof Hossain is Finance Director of BURO, Tangail. His main research interest is providing financial services for the poor.

Paul Mosley is Professor of Economics and Director of the University Development Centre at the University of Reading. His research interests include poverty reduction, structural adjustment and agricultural development. He has worked extensively in East Africa and regularly conducted research in South Asia and Latin America.

Rehman Sobhan began his working career on the faculty of Economics, Dhaka University. He served as Member, Bangladesh Planning Commission, Chairman, Research Director and Director General, BIDS, and as a Visiting Fellow, Queen Elizabeth House, Oxford. He was a Member, of the Advisory Council of the President of Bangladesh in charge of the Ministry of Planning and the Economic Relations Division. He is today the Executive Chairman, Centre for Policy Dialogue.

Rushidan I Rahman is a Senior Research Fellow at the Bangladesh Institute of Development Studies (BIDS), Dhaka. She also served as a lecturer in the Dept. of Economics, Jahangirnagar University. She obtained her Masters degree from Dhaka University and from the University of Sussex and her Ph.D. at the Australian National University. Her research interests include agricultural development, labour market and unemployment. Other factor markets in rural Bangladesh, micro-credit and gender issues related to labour and factor markets. She had published research monographs on 'Wage Employment Market for Rural Women' and 'Structural Adjustment and Health Care Services in Banlgadesh' and a large number of articles on the rural labour market and agriculture in Bangladesh.

Stuart Rutherford is from London but has been living in Bangladesh since 1984, where he worked first as the Country Director of ActionAid, an international NGO, and subsequently as a researcher, writer and practitioner of financial services for the poor. He is a Visiting Fellow at the Institute for Policy Development and Management at the University of Manchester. As well as his work in Bangladesh he visits India and Vietnam regularly where he assists in the development of financial services schemes. He is a Board Member of ASA, the Association for Social Advancement, a large Bangladeshi financial services NGO.

Syed M Hashemi received his Ph.D. from the University of California at Riverside in 1984. He has been teaching Economics at Jahangirnagar University in Banlgadesh since then. He is currently in charge of the Programme for Research on Poverty Alleviation at the Grameen Trust. The Programme intends to reorient research in Bangladesh away from donor driven priorities to the needs at the grassroots level. Professor Hashemi is also an activist with the Agricultural Wage Workers Union in Bangladesh.

Yuwa Hedrick-Wong is a development economist with over 16 years of international experience, spanning some 30 countries, encompassing the fields of rural development, health economics, institution capacity building, policy analysis, strategic planning, private sector initiatives. In the context of Bangladesh, he has been working with Proshika Manobik Unnayan Kendra's research, monitoring and evaluation of staff since 1992 to design and implement its comprehensive internal monitoring, impact assessment, and management information system. In addition, he has also provided training services in evaluation design, impact assessment techniques, and data management to Proshika's research staff. He is currently involved in the Health Economics Unit of the Population and Health Project in Bangladesh, tasked with policy research in key health issues.

Acknowledgements

The editors would like to thank various colleagues and friends who have assisted in producing this book to a very tight deadline in order to bring the evidence and critical analysis contained in these papers to the attention of those concerned with the contemporary debates about financing the poor in different parts of the world. Certainly the publishers should be acknowledged for recognising the urgency of the enterprise. The workshop (on which the papers were based) could not have occurred without the energies and enthusiasm of its organisers and hosts led by IDPAA, Proshika but crucially supported by representatives from the Credit and Development Forum (Sukhendra Kumar Sarkar, Chairman; Md.Yahiya, Director; Palsh Kumar Bagchi, Programme Officer) and the Bangladesh Institute of Bank Management (Dr.Toufic Ahmad Choudhury, Faculty Member and Dr.Bandana Shaha, Faculty Member). Additional support was given by David Hulme from the University of Manchester, who joined the steering committee. The support of UK-ODA and UNDP, Dhaka should also be acknowledged.

We would especially like to acknowledge the support throughout of Mr. Mahbubul Karim, the Head of IDPAA, who realised the significance of the issues and committed staff and resources, as well as chairing the planning sessions. Sharif is especially grateful for the enthusiastic advice and support of Stuart Rutherford and Graham Wright in developing ideas for the workshop. We are also grateful for the financial support provided by Proshika and the Ford Foundation for the preparation of the book including appointing an editorial assistant and enabling Wood and Sharif to meet in Princeton. The staff of IDPAA and elsewhere in Proshika once again revealed their professionalism and ethical commitments throughout the workshop. The authors of the papers in this volume have cooperated strongly with the editors both in preparing and presenting papers to the worksop, but also in making revisions quickly for this publication.

Moving to the preparation of the book itself, the editors have been very grateful indeed for the efficiency, initiatives and energies of Ms.Shukonya Shireen, the Administrator of IDPAA. Both of the editors left Bangladesh soon after the workshop to Bath (and a number of other countries) and Princeton. But the three of us have been in close touch by email as revised papers have been gathered in and negotiations with publishers commenced. In Bath, UK, we were fortunate to appoint Ms.Hazel Wallis as an editorial assistant. She is now quite experienced in working with Wood on Bangladesh material and was able to quickly assimilate the issues and objectives of this publication: organising a standard format throughout; proof reading and editorial corrections; editing Prof.Sobhan's speech into a draft for us to consider; and preparing final copy for publication. Mark Ellison, Administrator, Centre for Development Studies at Bath managed the contracts and logistics with his usual efficiency. Wood is particularly grateful to his colleagues (especially Allister McGregor and James Copestake) for advising on themes and key issues during the preparation for the workshop, as well as adjusting to interruptions in teaching. His graduate students have also been tolerant of the disruption to classes in late October. And his wife, Angela, has once again graciously accepted absences from the home both during the workshop and in finalising the book. Sharif would like to thank her parents for their encouragement and for tolerating her unsocial work schedule.

Princeton, October 1996   *Geoffrey D Wood*
                                                         Iffath A Sharif

Acronyms

ADB	Asian Development Bank
AKRSP	Aga Khan Rural Support Programme
ASA	Association for Social Advancement
ASCA	Accumulating Savings and Credit Association
BARD	Bangladesh Academy for Rural Development
BASIC	Bank of Small Industries and Commerce
BIBM	Bangladesh Institute of Bank Management
BIDS	Bangladesh Institute for Development Studies
BKB	Bangladesh Krishi Bank
BKK	Badan Kredit Kecamatan, Indonesia
BMET	Bureau of Manpower Employment and Training
BRAC	Bangladesh Rural Advancement Committee
BRDB	Bangladesh Rural Development Board
BRI	Bank Rakyat Indonesia Unit Desas
CDP	Centre for Policy Dialogue
CGAP	Consultative Group to Assist the Poorest
CIDA	Canadian I Development A
CO	Community Organizer
DTW	Deep Tube Well
EIG	Employment Income Generating
EPI	Expanded Programme on Immunization
GOB	Government of Bangladesh
HDI	Human Development Index

HRLE	Human Rights and Legal Education
HYV	High Yield Varieties
IDPAA	Institute for Development Policy Analysis and Advocacy (within Proshika)
IFAD	International Fund for Agricultural Development
IGA	Income Generating Activity
IGVGD	Income Generation for Vulnerable Group Development
IRDP	Integrated Rural Development Programme
KIE-ISP	Kenya Industrial Estates - Informal Sector Programme
KREP	Kenya Rural Enterprise Programme
KURK	Kredit Usaha Rakyat Kecil, Indonesia
MCI	Micro-Credit Institutions
MCS	Micro-Credit Summit
ME	Micro Enterprises
MFI	Micro Finance Institutions
MIS	Management Information System
MMF	Malawi Mudzi Fund
MSFSCIP	Marginal and Small Farms Crop Intensification Project
MTCP	Medium Term Credit Programme
NCB	Nationalized Commercial Bank
NFPE	Non-formal Primary Education
NGO	Non-governmental Organization
NTG	Non-target Group
ORT	Oral Rehydration Therapy
PEP	Productive Employment Project
PO	People's Organization

PRA	Participant Rural Appraisal
PVDO	Private Voluntary Development Organization
RAKUB	Rajshahi Krishi Unnayan Bank
RDP	Rural Development Programme
RESP	Rural Employment Sector Programme
ROSCA	Rotating Savings and Credit Association
RRA	Rapid Rural Appraisal
RRBs	Regional Rural Banks, India
SAARC	South Asian Association for Regional Co-operation
SACA	Smallholder Agricultural Credit Administration, Malawi
SANASA	Thrift and Credit Co-operatives, Sri Lanka
SHG	Self-help Group
TCCA	*Thana* Central Co-operative Association
TG	Target Group
TRDEP	*Thana* Resource Development and Employment Programme, Bangladesh
UK ODA	United Kingdom Overseas Development Administration
UM	Unit Manager
VO	Village Organization
VOCP	Village Organization Credit Programme

Glossary

Aman	Main rain fed, monsoon rice crop
Babsha wallah	Entrepreneur
Baki	Credit
Bari	Village
Bhoy	Fear
Bittahin	Landless
Bustee	Slum
Crore	10 million
Haat	Local market
Hilsha	Type of fish
Kendra	Centre
Khas	Government-owned resource (land, forest, pond etc)
Khela Samity	Game club (names used)
Loteri Samity	(for ROSCAs)
Kisti	Repayment
Kormi	Development Worker
Lojja	Shame
Maund	Unit of weight (approx 40 kg)
Mohajan	Moneylender
Motha	Roots (of trees)
Rin	Credit
Ruti	Bread

Samity	Society
Shalish	Court (of justice)
Shonchoi	Savings
Taka/Tk	Bangladesh unit of currency (US $= Tk 40)
Thana	District
Tin bella bhat	Three meals a day
Zila	District

INTRODUCTION

Geoffrey D. Wood and Iffath Sharif

Context: Micro-Credit Summit

This volume of papers on poverty and finance in Bangladesh arose from a workshop with the same title held in Dhaka, Bangladesh in August 1996. That workshop was conceived a couple of years earlier before plans for the Micro-Credit Summit (MCS) had been announced. This summit in Washington in February 1997 has now become an important focal point for all current discussions on poverty and finance in different parts of the world, and constitutes a major rationale for converting the proceedings of the Bangladesh workshop into a volume to appear in time for the summit. The editors and the Institute for Development Policy Analysis and Advocacy (IDPAA), Proshika in Dhaka convened the workshop (with the Credit and Development Forum, and the Bangladesh Institute of Bank Management as co-sponsors) and consider that the arguments contained in the following papers have a special relevance to the summit. It is no secret that while there are numerous models and practices around the world linking financial services to the livelihood strategies of the poor, the experience in Bangladesh has become globally influential, especially through the record of the Grameen Bank in currently lending to approximately two million families. To the extent that the Grameen Bank, through its research wing the Grameen Trust, is sharing, indeed promoting, its 'Bangladesh' model for wider application worldwide through the MCS, then it is also relevant to share the self-critical analyses from both the Grameen Bank and other micro-credit providers in Bangladesh together with commentaries by other observers with long experience of poverty-focused rural development in the country.

Although obviously not alone in the global landscape of poverty-focused micro-credit, Bangladesh has become the site of intensive experimentation reflecting a particular combination of conjunctural forces: a liberation-derived commitment to the poor among a generation of recent graduates in the early 70s; coinciding with post-liberation, military repression of progressive open politics which diverted many of these graduates into the formation of NGOs, supported initially by a small group of 'like-minded' progressive donors (e.g. Scandinavia, Canada and the Netherlands) and some Northern NGOs like OXFAM and NOVIB; and a subsequent widening of this donor support in the late 80s as both the Bangladesh state continued to reveal its limitations and the donors' neo-liberal dogma in the context of mass poverty looked increasingly over-optimistic, even immoral. Thus although there are many key micro-credit examples elsewhere in the world, e.g. in Indonesia, Kenya and Bolivia (Hulme and Mosley 1996), Bangladesh has provided models of recognised global significance, and is most famous for the Grameen Bank model.

With its funding shared between its borrowers and the government of Bangladesh, this Bank cannot be classified as an NGO, but it arose as a model from the same conjuncture of forces noted above. Professor Yunus (its founder and current Chairman) was an economics professor at Chittagong University after liberation, but unlike other professors he was in touch with rural development experimentation occurring both within the early NGOs and post-Comilla, government initiatives. He was also alert to the general conditions of poverty, especially on his own 'back doorstep' in the villages around the Chittagong campus. He gained direct understanding of the liquidity problems of the poor, their entrapment in usurious relations when trying to overcome the endemic problem faced by the poor (namely, extreme fluctuations and unpredictability in the flow of income), and their lack of access to the formal banking sector. He 'spotted' that this sector could not contemplate the risk entailed in lending without collateral nor the costs involved in gaining information about borrowers' reliability as an alternative risk avoidance strategy. The now familiar joint liability model arose from this awareness, since it offered a method for overcoming information costs and securing enforcement within the same institutional formula, thus contributing to the sustainability of the lending agency through reducing its transactions costs and ensuring frequent and regular repayment of loans. It is this model that has been able to move to scale (more than 25% of the total estimated global portfolio of micro-lending to the poor), demonstrating a systemic replicability virtually independent of specific culture and social structure. It is in this way that the Bangladesh experience has become so relevant to the deliberations of the MCS.

Unfortunately, the world has heard less about the other experiences arising from the post-liberation conjuncture of forces in Bangladesh which set limits to the validity of the Grameen Bank model and takes us into both a broader financial services agenda and other social development strategies for poverty removal. Given its focus, this volume of papers concentrates more upon the 'broader financial services agenda' but it will inevitably be making connections to other parts of the poverty removal agenda.

Before referring in more detail to that broader agenda, it is important to reflect carefully upon the status of 'criticism' of the strengths and weaknesses of various models and approaches against different, often conflicting, sets of objectives. It will be obvious that one of the motivations behind the workshop and the publication of this volume of papers is to point out the limitations of a narrowly conceived and practised strategy of micro-credit intervention in poor people's lives for achieving objectives of poverty removal (in contrast to poverty reduction or poverty alleviation). Such a motivation therefore involves a criticism (*inter alia*) of the Grameen Bank model, of its replicas and of the interests of those sets of actors (especially among the donor community) who appear keen to exercise their influence and direct their resources towards a monotheistic micro-credit formula based on the model and its variants. The model is developing 'panacea' status, thus inevitably becoming dangerous for the over-reductionism involved. We have to be alert to the evangelism entailed in the promotion of the panacea (Rogaly 1996, and Copestake 1995), in the same way that others object to mono-cultural tendencies in other sectors (whether cereals, livestock breeds or project management rhetoric).

However, the editors are very clear themselves and wish others to be similarly clear that the criticism entailed in this volume is intended to move all of us forward in considering appropriate strategies for overcoming poverty and contributing to sustainable poor people's livelihoods. That intention needs, then, to recognise and acknowledge the enormous contribution made by the Grameen Bank to shifting the discourse about poverty and finance. In almost a Kuhnian way, the Bank's small group, joint liability model (premised upon the assumption that 'borrowers know best' with repayments that were explicitly to come from poor people's propensity to save rather than from credit supported, dedicated project activity) has represented a paradigm shift even if its separate ingredients were scattered across pre-existing small scale credit programmes. The Grameen Bank uniquely brought these different elements together and then moved to a scale which demonstrates their combined, systemic viability.

However, this viability is being defined within a narrow, perhaps under-ambitious, set of objectives concerning the relationship between the regularity and frequency of repayment and the sustainability of the lending institution (where costs of operation can be adequately recovered from a large number of small scale loans to large numbers of poor people—especially women, the most weakly positioned category of social actor in Bangladesh society). It is therefore appropriate to ask, alongside others both within the Grameen Bank and outside it, to what extent the model as currently practised can address more ambitious objectives of poverty removal, implying an irreversible shift in the condition of the poor (e.g. vulnerability to shocks and major life-cycle expenses, control over key means of production, wage and rent bargaining, access to human capacity building services such as health care and education, price-making in markets, and societal respect). It is also appropriate to ask as Jain does (1996) whether the successes of the Bank are attributed to its joint liability, social collateral model (with the key pay-off of reduction in transactions costs) or to the efficacy of its local level management in enforcement, employing 12000 staff. If the latter, then this implies a severe constraint to the scale ambitions of the MCS (i.e. micro-credit to 100 million people by the year 2005). It is also appropriate to ask whether the joint liability strategy involving small groups of women confines the use of such credit to small-scale, low turnover activity which is essentially non-threatening to the male- and class-dominated local political economy; and whether this restricts the effect of such credit to an important welfare function (poverty alleviation) rather than irreversible structural change. It is also appropriate to ask whether the emphasis upon micro-credit and the seductiveness of this emphasis to donors (either as a target for aid resources or as an alibi for overall aid reductions) is crowding out other essential elements of the poverty removal agenda.

It seems clear from reports of the Prepcom meeting for the MCS that the spirit of renaming the summit a Micro-Finance Summit had broad approval even though an actual re-naming was resisted, presumably due to organisational difficulties at this stage. The reactions to the original pronouncements from the MCS initiators appear, then, to have already shifted the terms of debate away from the 'mono-culture of micro-credit' towards a broader conception of financial services. No doubt this process has been assisted by recent published contributions from Rogaly (1996) and Hulme and Mosley (1996), and the critical works to which they have referred (such as Copestake 1995). In the Bangladesh and South Asia regional context, the persistent arguments of Rutherford and Wright over the last few years concerning savings and access to them have been consolidated into papers for this volume. Although the concern for 'broader financial services' is itself seductive, partly because it persists with the assumption that 'borrowers know best', it would be a shame if such a preoccupation displaced other questions about the relationship between credit and poverty removal, and therefore about other models and objectives embodied in lending to the poor. The 'broader financial services' agenda appears to embody the idea of poor people as 'customers' as in rich countries, requiring a wider range of such services: insurance, mortgages, overdrafts, pensions, non-specific credit (as in plastic cards), loan support to business plans, investment services as well as savings (with variable access/tax incentives). While such a conception may be attractive as a long term goal within a vision of a capitalised, computerised and market integrated Bangladesh, as an operational conception of the present target group it remains wide of the mark in the short and medium term.

It is therefore important to guard against an extremist new 'new wave' of financial services advocacy (Rogaly refers to the 'new wave' evangelists for micro-credit) which over-fantasises the structural position and capacities of the poor to resolve their poverty through a more complex portfolio of financial services, while ignoring the realities of the political economy and culture. It is important to remember that the beauty of the Grameen Bank model lies in the extent to which it can work within limited objectives almost independent of political economy and culture. Replacement panaceas would have to enjoy similar autonomy and elegance, whereas moving to the broader financial services agenda risks assuming too much about the facilitating conditions for such a shift within markets, organisational arrangements, financial 'technologies', and new forms of morality and trust (e.g. in local ROSCAs).

One key problem in moving towards the broader financial services agenda on any significant scale will be the adaptability of existing credit delivery NGOs along several dimensions. The most obvious adaptation problem will be opening access to savings where they are deemed by NGO and donors alike as essential collateral to ensure lender sustainability, as is very clear, for example for BRAC in Bangladesh and AKRSP in Pakistan. NGOs which currently allow access to savings such as Proshika are regarded with suspicion by donors, who lack faith in an over-reliance on revolving payments and interest earnings. But NGOs would have further adaptation problems in switching their mentality from beneficiaries to customers, since this would threaten their strategic control over the link between resources and objectives and perhaps weaken broader social leverage in both advocacy and campaigns (e.g. joint NGO-People's Organisations attempts to bring private and government property into wider forms of common property and management). In short they would have to function more like reactive banks than proactive strategists where currently their credit-based entry into local livelihood struggles offers legitimacy and leverage to pursue more challenging social development, arguably more structurally significant for long term, sustained poverty removal even if risky and uncomfortable for their poor beneficiaries in the short term. Another way of cautioning about the new 'new wave' of micro-finance advocates is to suggest that the language of 'customers', while populist in the sense of 'customer knows best', promotes individualism (or small group-ism at best) when poverty removal (in contrast to amelioration) requires wider forms of organisational solidarity.

These concerns lead us back to the central question of the place of credit and finance in the quest for poverty removal. They tell us that while the micro-credit agenda is too restrictive and failing to address the wider financial needs of the poor (as reflected in their own initiatives and described especially by Rutherford in this volume), we should accept that both credit and broader financial services have their place as a necessary but not sufficient condition in pursuit of poverty removal. They tell us that the protection of the structural value of poor people's financial assets requires wider forms of mobilisation and intervention in which poor people need to be more than just customers and NGOs need to be more than just MCIs or even MFIs if the agency of the poor is to be promoted. In this way, poverty removal always has to be context specific. These were our concerns for convening a workshop to review the experience of different approaches and models in Bangladesh even before the announcement of the MCS. We return to these issues in the concluding chapter to this volume in which we summarise the preoccupations of the workshop with the relation between micro-credit and micro-finance, and indicate the case for the broader social development agenda as a precondition for securing borrower sustainability.

Workshop on Poverty and Finance in Bangladesh

The proposal for IDPAA to convene a workshop comparing the experiences of different models and approaches in lending to the poor arose from Wood's reactions to appraising the Action Aid programme in Bhola island in July 1994. This programme had sought explicitly to reproduce a Grameen Bank type approach to supporting the livelihoods of poor people in the southernmost part of the island. These communities remain among the most vulnerable and isolated in the country, and Action Aid was certainly to be applauded in locating its first operational programme in Bangladesh here. However the appraisal team, led by Wood, developed a strong consensus that the intervention was too narrowly conceived, and that many of the potential benefits from the work of enthusiastic and dedicated personnel were being lost (Wood et al 1994). The programme prided itself on focusing upon female borrowers and limiting its contact with them to regular meetings in which savings, repayments and credit disbursal was managed. There was minimal attention to mobilisation and intervention beyond this, and little understanding of the household and community dynamics through which the value of these loans were refracted. The potential structural value of such lending appeared to require protection from the political economy and cultural forces in the localities. There was also concern that such loans were encouraging poor people to enter already over-crowded petty processing and trading markets. There was even concern that the focus upon female borrowers was contributing to an inflation of dowry expectations in the locality, placing further pressure upon other poor families not participating in the scheme.

Secondly, even with such reservations about 'credit only' approaches, how could a credit provision for the poor be sustained in the area after the departure of Action Aid? The answer prompted much debate about institutional options, but a self-provider option through a federated structure of mobilised groups seemed to be the weakest precisely on account of the limited, credit-oriented, mobilisation which had guided the programme so far. The whole exercise prompted a series of questions about securing the sustainability of borrowers and lenders in poverty-focused development. The outcome of this appraisal exercise was to reflect further upon the relationship between credit intervention and the wider social development activities of NGOs in their work with the poor of Bangladesh, and to propose that the time had come when NGOs and the Grameen Bank, together with official programmes such as the RPP in BRDP, should share their models and autocritiques in order to identify more clearly what could be achieved via the different models.

Wood therefore posed the following questions to colleagues in IDPAA. To quote from his original fax:

> "It seems to me that there is a large debate simmering about credit options and dilemmas within NGOs, GOB, banks and donors in Bangladesh. Should it be connected to mobilisation efforts or kept separate? Should NGOs expand their capacities and substitute for the banking system in the rural sector and the informal urban sector?

Should NGOs act as intermediaries between people and banks? How can small NGOs resolve the credit problem for themselves or must they attach themselves to a larger credit provider? Can people easily belong to (be loyal to) one organisation for a 'mobilisation' agenda, and yet be attached to another for credit purposes? Is there a case for an NGO bank, especially to resolve the problems of smaller NGOs? What are the prospects for mature groups/people's organisations to operate self-managed savings and credit schemes (ROSCAs)? Can mature groups be graduated to special banks? What package of financial services are actually required by the poor? What are the problems of collective credit behaviour versus strategies for individual families? What are the gender implications of various forms of credit intervention? What is the impact upon other forms of rural indebtedness? How does credit affect local market behaviour e.g. inflating the prices of certain assets, commodities and exchanges (like dowry)?"

A further context for such an exercise were two conferences on micro-credit held during early 1995 in Reading, UK and also in Dhaka, Bangladesh in which donors and the World Bank played a prominent role. There was a strong feeling among the NGO participants and other observers that the World Bank and key donors like USAID and UK-ODA were keen to push multi-sectoral, social development-oriented NGOs into the narrower function of micro-credit institutions (MCIs) and eventually micro-finance institutions (MFIs). The premise behind such influence was that as NGOs increased in scale of operation and significance, so their ability to sustain costly social development activity at existing levels of staff intensity was unsustainable. If NGOs wished to continue to be attractive to donors at a larger scale of activity, then they had to show that they were sustainable as institutions in the longer term by securing cost recovery through micro-credit lending and other financial services. Furthermore, these donors based this analysis upon an emerging conventional wisdom that the Grameen Bank had proved the positive impact of credit upon the poor without the costly accompaniment of wider social mobilisation around common property access, wages and rent struggles, and so on. This picture seemed to be reinforced by BRAC's reduction of its original social mobilisation agenda, and by selective examples from the Hulme and Mosley studies elsewhere.

Such a policy carries profound implications. It offers an alibi for donors as well as nervous host governments to withdraw from a broader analysis of poverty into a narrower, neo-liberal, conception based on poor people's financial liquidity. It turns its back upon decades of structural analysis about the social and economic inequalities embedded in transforming political economies, and adopts panacea labelling of the poverty problem in ways which are non-threatening to classes in dominance (globally and nationally). These concerns are reinforced by the preparation discussions for the upcoming Micro-Credit Summit in Washington, USA in February 1997; as well as the continuing endorsement of NGOs as MCIs by The World Bank, the Clinton Administration (with the support of the South Shore Bank in Chicago), and neo-liberal academics (for a review of such positions, see e.g. Hulme and Mosley 1996, or more briefly Matin in this volume). These neo-liberal academics are increasingly restricting the function and purpose of any group formation in poverty-focused development to the policing of micro-credit loans in order to reduce transactions and information costs (in particular the 'Information School'). It has clearly become important to react to and analyse these trends.

This policy pressure from key donors polarised the debate into a concern for the sustainability of the lenders on the one hand, and for the viability of the borrowers on the other. Since the donors were in direct contractual relationships with NGOs holding large micro-credit portfolios (e.g. BRAC and Proshika in Bangladesh), their primary concern for the sustainability of these lending institutions was understandable. But their concern went beyond this to the prospect of gaining further 'value for money' on their grants and soft-lending by concentrating on those activities for which direct cost recovery could be demonstrated. Social development was in danger of being regarded as an optional tax on lending, especially as social development impact was so difficult to measure objectively.

Clearly Proshika in Bangladesh felt especially vulnerable to this analysis. It has a large RLF alongside a continuing commitment to its original mission of conscientisation and mobilisation among groups of the poor in order that they can struggle more successfully for rights and entitlements in the political economy, and enter commodity and labour markets on more even terms with other classes of social actor. Furthermore, Proshika was resisting pressure to reduce client access to their savings in order to retain these savings as collateral against a further expansion of its credit portfolio. While acknowledging the validity of the donor preoccupation with lender sustainability, Proshika was concerned that the issue of borrower sustainability was being reduced to a lowest common denominator of 'ability to repay' which was regarded as an indicator of securer income and a propensity to save. For Proshika, at least, this was not a sufficient indicator of borrower sustainability.

The outcome of such exchanges was that IDPAA, Proshika and UK-ODA were keen to convene a poverty and finance workshop on the Bangladesh experience across a range of models, but the former wanted to emphasise the circumstances of the borrower and the latter the lender. The resolution was two workshops to reflect these different emphases, although, of course, everyone concerned knows that borrower and lender sustainability are mutually entwined. Indeed since the whole micro-credit discourse stems from the reluctance of commercial banks to undertake the risk of lending to the poor due to perceived threats to their own commercial viability, then a proof of viability in lending to the poor is certainly required. However if that viability is to be achieved at the expense of borrower sustainability by restricting the mobilisation agenda, then the outcomes for the poor may only consist of a modest contribution to short term welfare and the prospect of long term graduation denied. Furthermore, such a preoccupation with lender viability may function to screen out the poorest for the same reasons as the commercial sector seek to exclude the poor altogether. Thus, in IDPAA, we considered that the workshop on borrower sustainability should logically precede a sequel on lender sustainability.

The second workshop on lender sustainability is scheduled for mid-1997 and will not only focus upon the viability characteristics of a 'hierarchy' or continuum of options from minimalist credit extension to a wide social development perspective which includes a credit dimension. Crucially it will also be focused upon the other dimension of lender sustainability in the heavily aided context of Bangladesh—namely accessing domestic capital markets for on-lending to the poor as part of the longer term strategy of reducing dependence upon donor capital. This second focus obviously returns us to the original problem of linking commercial capital to the interests of the poor. However the presumption here is that NGOs (either severally or collectively) will act as mediators and brokers for such capital rather than seek again the direct involvement of the commercial banking sector in lending to the poor. Nevertheless, such an agenda continues to request that the commercial capital sector put a proportion of its funds at the disposal of such intermediaries for a market return on capital lent. NGOs are therefore obliged to convince themselves that they can square the circle of lending in a sustainable way to the poor in a manner which delivers long term sustainability for the poor themselves, while delivering at the same time adequate return on capital to Bangladesh's capital markets.

This second workshop will therefore also have to include a focus upon regulation and supervision by government or government approved authorities upon the transactions between domestic capital markets and NGOs, which will also entail closer examination of changes in the status of NGOs as financial institutions in terms of the rights of borrowers and depositors. It is important to recognise that interlinkage between these formal and informal financial sectors will entail flows of funds in both directions, with the savings of the poor also placed at the disposal of capital market investors to earn returns for the poor and their NGO brokers. Under such conditions, savings are exposed to higher risk which will require policy on guarantees and insurance, as well as regular monitoring of risk spread in such savings portfolios and the performance of brokers in seeking a balance between best and secure returns.

Borrower Sustainability: Introducing the Papers

As indicated above, the workshop from which these papers are derived was designed to focus mainly upon borrower sustainability, while recognising a certain artificiality in the separation from lender sustainability. Quite properly, therefore, the boundary between the two sets of themes is not always maintained in the papers and nor was it in the workshop discussions, as reflected in the concluding chapter to this volume.

A brief comment on the organisation of the workshop should assist the reader on the rationale behind this combination of papers and thereby offer guidance on selecting items within this volume. There was always the intention of bringing together a small group of highly informed practitioners, innovators, critical observers and analytic donors to debate issues away from the glare of a large gathering which might prompt defensiveness of cherished models and practices rather than honest self-criticism as the basis for moving forward. To a considerable extent, this was achieved with most of the focused sessions consisting of 30-40 participants. At the same time, the workshop was an advocacy event in the sense of seeking publicity for the issues raised as part of contemporary policy formulation within the new government, and among donors reacting to that incoming government.

Advocacy events do entail some larger scale occasions. Their legitimacy is enhanced by having several sponsors, so that IDPAA was joined from the outset by the Credit and Development Forum (a networking NGO) and the Bangladesh Institute of Bank Management, as well as UNDP and the UK-ODA. The steering committee of the workshop also included Wood from the University of Bath, and Hulme from the University of Manchester, both in the UK, working in close collaboration with IDPAA and Ms. Iffath Sharif, one of its research associates. The significance of these occasions is also reinforced by the presence of key political and policy figures making their observations on the theme. The workshop was therefore opened with a large gathering, addressed by the Finance Minister; there was an open panel session at the end of the second day with bankers, professors, heads of large NGOs and a Member of Parliament; on the third day, the editors of this volume together with the co-sponsors summarised the main issues arising from the papers and discussions (see the concluding chapter for an expanded version of this summary); and the summary was presented to a closing large session (approximately 500 participants present, including the national press) on the fourth day, before inviting reactions from the floor and ending with a closing set of reflections by Prof. Rehman Sobhan, ex Director-General of the Bangladesh Institute of Development Studies, an early post-liberation planner, co-ordinator of the

Task Force which presented a multi-sectoral policy analysis during the first interim government in early 1991, and presently head of the Centre for Policy Dialogue—a think tank which now produces an annual country economic memorandum as an alternative to that of the World Bank. Prof. Sobhan's address has been edited by us (with his permission, of course) and appears as Chapter 4 in Part I of the volume.

There was always the intention to combine issue papers with 'autocritique' case-studies of different credit and micro-finance approaches in Bangladesh, which together would represent the range of options from large-scale, directed micro-credit services to small-scale, experimental, even involving local indigenous initiatives. We were also able to maintain a comparative dimension through the adapted and summary presentation by David Hulme of the recent Hulme and Mosley study of 'Finance Against Poverty' (1996). This is how Mosley appears in this volume as a co-author, while not being present at the workshop. It is expected that the second workshop will draw more explicitly on the comparative experience of lender models to inform appropriate policy reform in Bangladesh (especially in regulation and supervision, which is a particular concern of the Bangladesh Institute for Bank Management, a co-sponsor).

Readers can see from the combination of papers in this volume that these intentions were broadly achieved. The organisers, and especially the editors of this volume, would like to place on record their appreciation for the co-operation of different institutional actors in contributing their self-critical analysis. This co-operation has continued into the revisions of papers under tight time constraints for inclusion in this volume—hence the particular gratitude of the editors. In the following review of the papers by the editors, there is summary description of paper contents interspersed with additional commentary and cross-referencing statements. We hope that the authors will not object too strongly to our embellishments and interpretations of the significance of their descriptive and theoretical analysis.

In organising the papers for this volume, the editors have been concerned to avoid a 'proceedings' formula of simply following the programme of presentations and discussions. Indeed, we do not include records of discussion as separate items in the structure of contents. Instead, the editors were concerned: to group the papers in five parts for clear thematic purpose; to encourage authors to incorporate relevant contributions from discussants in the revisions of their papers; and to develop the overall argument of the workshop in a way which offers clear messages for the MCS and beyond. In brief, the message is that Bangladesh does not represent a single formula of poverty-focused micro-credit for the rest of the world to follow, but rather a proliferation of strategies ranging from official government, semi-official government (i.e. Grameen Bank), NGOs (with many internal variants of emphasis) to very local, indigenous grassroots experimentation. Embodied within this overall argument is the proposition that the preoccupation with credit and financial services for the poor should not displace wider strategies for poverty alleviation, reduction and removal; and furthermore, that these wider strategies are a precondition for realising the full potential value of credit and financial interventions. The livelihoods of the poor cannot be sectoralised and addressed on one dimension only. This seems to be the conclusion shared by most of our contributing authors, and is certainly consistent with the comparative findings of Hulme and Mosley (1996) as well as the general thrust of Rogaly's concerns about 'micro-finance evangelism' (1996).

There are 16 papers arranged into five parts: Overview; Case-studies and Themes; Problems of Reaching the Poorest; Micro-Credit and Limitations of Scale; Micro-Credit—A Restricted Approach to Financial Services. These parts are followed by a Conclusion which expands the summary prepared for the final session of workshop by the editors and co-sponsors, and which tries to develop a clear vocabulary for distinguishing three strategic options: credit alone; credit plus; and credit with social development. This threefold distinction reminds us that one of the key institutional problems to be resolved is how development NGOs with strong mobilisation agendas can retain those agendas while seeking their own sustainability through the management of other people's money: the theme of the sequel workshop.

Overview

The four papers in Part I (Overview) perform an agenda-setting function. The paper by Iffath Sharif was prepared over several months as a background paper to the workshop as well as a policy briefing paper for IDPAA. In its various stages, the paper was deployed as a negotiating document in developing support for the workshop. In this sense, it reviews the broad institutional landscape within Bangladesh around the poverty and finance nexus, highlighting the problems of borrower and lender sustainability and the interconnections between them. The paper is conceptually organised around the theme of graduation for both borrowers and lenders. After a section which deconstructs 'the poor' into constituent categories of extreme poor, moderate poor and vulnerable non-poor, she argues that graduation from all three categories requires different kinds of support and services—i.e. flexible financial services as well as other forms of intervention. However she is also keen to point out the graduation problems of micro-credit institutions (MCIs) in their search for sustainability via access to additional commercial funds to increase credit coverage. To the extent that the future of these MCIs has to consist of integration into the formal financial sector, then substantial policy changes will be required not only within NGO/MCIs, but also between them where co-operation is required, and with formal sector banks and the government (in terms of regulation and supervision). This constitutes the new policy agenda, and a number of key institutional issues and possibilities are reviewed. Although Ms. Sharif has been concerned that this paper stretches across both issues of lender as well as borrower sustainability, while the first workshop and this volume focuses upon the latter, we have decided that the overview function of the paper in posing the trade-off problem of poverty reduction and financial sustainability via the integrating concept of graduation offers an efficient insight into the wider policy context and the dilemmas to be faced as the agenda is moved forward.

Greeley's paper focuses upon impact analysis as the crucial test of the efficacy of micro-credit interventions by returning us to the issue of poverty measurement. This is a timely discussion in the light of Rogaly's observation (1996 p 103) that analysts have turned their back on impact assessment referring to Mansell-Carstens' critique of income and well-being measurements (1995) in favour of simpler viability indicators such as repayment rates, subsidy-dependence index and numbers of borrowers and savers. There is concern that the MCS will also ignore impact assessment. Greeley decides to cut through these problems by taking us back to the use of a poverty line based on food consumption. While recognising the developments in participatory appraisal of poverty (meanings and measurements), the ethical theorising of Rawls (1971), Doyal and Gough (1993) and Sen (1989) *inter alia*, and other attempts to pluralise the concept of poverty (as in the UNDP's HDIs), Greeley points out that any concern for borrower sustainability has to be rooted in an individual's food security. When all the subtlety is stripped away, a food-based poverty line in a society like Bangladesh remains a valid indicator of priority needs for the poor, and it has the virtue of being measurable. He then proceeds to distinguish between levels of poverty and correlates of poverty: the former establish the priorities for intervention, the latter guide the precise forms of intervention required. He continues by arguing that not all poor households have been able to benefit from micro-credit programmes: especially, of course, drop-outs and non-members of such programmes. His concern is that in the rush for micro credit panaceas such categories of the poor will be by-passed as other targeted resources become increasingly diverted to the micro-credit formula. He concludes by noting the World Bank's threefold poverty discourse of labour intensive growth, human resource development and safety nets with micro credit particularly directed towards the first of these. However the problem here is the extent to which the economic activities promoted by such micro-credit interventions are concentrated on non-tradable output with little prospect of enterprise graduation—a theme to which later papers return. Finally he also touches upon the persistent theme of this volume: namely whether the concern for the financial viability of the lender will undermine the present compassionate ethics of development NGOs, and further exclude the poorest.

The Hulme and Mosley paper, as a condensation of some of their arguments for their 'Finance against Poverty' volume perhaps requires less introduction to the readership of this volume. The chapter examines the contribution that micro-credit and micro-finance can make to the alleviation and removal of poverty, drawing on the comparative materials from their study of 13 financial institutions in seven countries. As noted earlier, the chapter therefore provides this volume with a comparative analysis within which to locate the Bangladesh cases and themes. Overall, the paper supports a broadening of the agenda from micro-credit to micro-financial services, but given problems of exclusion, other forms of intervention should not be crowded out by a new orthodoxy. As with Greeley, the paper opens with a conceptual discussion of poverty in order to identify the criteria to be deployed in assessing poverty impacts. The argument which follows is then organised around the relation between financial services and the three themes of poverty (or income poverty), vulnerability and deprivation. The key 'income poverty' finding is that income-generating credits are not scale neutral and have differential utilities and effects for the differentiated poor. From this, they conclude that credit schemes are most likely to benefit the incomes of the middle and upper poor, with the poorest or 'core poor' therefore requiring other forms of assistance. And secondly, that institutions seeking their own financial viability through income-generating credit to the poor will have a tendency to concentrate upon the middle and upper poor. The paper then asks what contribution is made by their case study agencies to reducing downward mobility pressures (structural, crisis/contingency and life cycle) and strengthening household security to overcome vulnerability. Various interventions are surveyed and evaluated, and the authors commit to an overall assessment that none of the case study institutions generates sufficient income for more than few borrowers to amass an asset base to provide pensions when their enterprise days are over. Micro-entrepreneurs, it seems, cannot provide for their old age, thus setting a firm limit to the notion of borrower sustainability via the panacea of micro-enterprise. The third, deprivation, theme outlines wider, non-material concepts of deprivation (including, crucially, gender issues) and the wider forms of intervention required to address them. The paper then concludes with a severe attack upon the search for universal models which fail to differentiate the poor and their needs, upon the over-inflated claims made for credit schemes as vehicles for social mobilisation, and the apparent abandonment by many agencies of their earlier attempts to create solidarity among the poor as the route to structural change in the political economy.

The reader should now be in a position to appreciate why the editors were keen to include, at this point in the volume, an edited version of the concluding address by Rehman Sobhan which we have titled 'The Political Economy of Micro Credit'. With an established preference for Marxian forms of economic and political analysis and a history of persistent critique of neo-colonialism and the dependency of Bangladesh (where NGOs *inter alia* are not spared), Sobhan directs his fire upon the structural conditions which reproduce poverty at both micro and macro levels, including the position of Bangladesh in the global economy which constrains as well as creates opportunities for the employment of the poor. He is looking for evidence of a connection between micro credit and sustainable transformation in the lives of beneficiaries. He is concerned about the exclusion of the very poor. He raises the question of causal relationship, if any, between 25 years of micro-credit and economic growth, noting that some households may have improved but that the overall incidence of poverty remains high even in the villages receiving micro-credit. He finds little attention either on the significance of market imperfections for undermining the value of credit to the poor, or on credit supported forms of entry into the local or national economy which might reorganise the basic structure of the economy more in favour of the poor, although he does note the attempts to capture water markets by the poor. He also teases the workshop participants for the shifts in vocabulary from mobilisation to social development, from micro credit to the 'up-market' concept of financial services, then moves on to the paucity of thinking about developing macro financial instruments to secure higher returns on poor people's savings. He criticises the over-dependence of NGOs on external funds instead of looking to domestic financial markets for their institutional sustainability. He concludes by reflecting on the problem of 'system-loss' in the present aid system, with an implicit concern that the MCS might be seeking to promote more aid flows behind a micro-credit initiative with accompanying system loss.

Case Studies and Themes

There are five papers in Part II (Case Studies and Themes) which describe and explore the limitations of different programmes by authors from those organisations and sectors. Represented are: the Grameen Bank (by S.M. Hashemi); among the NGOs, Proshika (by Hedrick-Wong et al), BRAC (by Chowdhury and Alam), and the Association for Social Advancement—ASA (by A.K. Aminur Rashid); and the public sector banks (in a composite paper by M.H. Khan). In terms of scale, these organisations together constitute overwhelmingly the bulk of lending to the poor in Bangladesh, as can be seen from the evidence provided in the papers. This is the rationale for placing these case studies alongside each other. However, it is important to note that there are large numbers of smaller organisations in Bangladesh involved in credit for the poor but on a much smaller scale. Some of these represent important experimentation with alternative models of wider financial services and are therefore placed separately in Part V as a 'critique in practice' of the narrower micro credit approaches.

These 'large-scale' case studies appear at this point in the volume since they also contribute to an agenda setting of themes which are then pursued in the subsequent Parts. The 'Proshika' paper by two outsiders to the organisation (Hedrick-Wong and Bosse Kramsjo) together with A.A. Sabri from IDPAA develops an opening argument about the connection between credit and social development, then illustrates some of these issues via a series of case studies from Kramsjo's 'rural rides' over the last two years before concluding with some policy challenges for the future. Proshika unashamedly espouses a generic model which asserts the interdependence of social and credit services, and seeks in this paper to demonstrate the logic both theoretically and empirically. It proceeds from the standard 'market failure' argument about the limits of the formal credit system in serving the mass demand from the poor for capital: imperfect information and imperfect enforcement. The response to these market failures has been via some form of social services featuring group formation to overcome the information and enforcement constraints. However the widening range of service configuration among the different models represent different assumptions about how much social development service is optimal. The Proshika 'position' at the intensive end of the social development (or social mobilisation) continuum is maintained not just as a response to the market failure problem, but also to realise the full value of borrowing in terms of successful participation in the political economy. The paper seeks to establish this argument empirically by reference to an impact survey across a sample of Proshika groups before presenting 4 illustrative case studies which reveal different 'performance' outcomes but which all show the case for social empowerment to accompany any credit or financial services intervention.

The paper on BRAC by Chowdhury and Alam summarises the BRAC approach, laying strong emphasis upon a holistic approach via social mobilisation for poverty alleviation. We learn from this paper that BRAC still maintains a public commitment to such holism despite a recent and widespread criticism that BRAC has retreated from its earlier social mobilisation stance into a narrower contemporary conception of micro credit delivery as part of a deliberate plan of lender sustainability. It is clear from this paper that BRAC is wrestling with this problem of how to reconcile its original analysis of what Greeley refers to as poverty correlates (which prompt a range of social services interventions) with its own sustainability that does not rely upon infinite grant support from its consortium of donors. Its frustrated plans to convert itself into a Bank (encouraged by its donors) represents the process of development NGO conversion into an MFI, as pursued by some global stakeholders at the 1995 Reading conference (and presumably in the upcoming MCS). The paper notes that BRAC has other contributory routes to sustainability via its franchise contracts with the state as in non-formal primary education or primary health promotion. After summarising its impact along a range of indicators, the paper concludes with a series of 'lessons learned'. Its commitment to a holistic approach, addressing a wide range of material and non-material poverty correlates is reiterated; credit is therefore identified as crucial but not sufficient; and credit is needed to support technology access for the poor to enable them to make a wider contribution to employment generation and economic growth through forward and backward linkages. However the problem of exclusion of the

48 *Who Needs Credit?*

'ultra-poor' remains, confirming the Hulme and Mosley finding that the poor gain less leverage from their credit and remain less attractive to a lender organisation seeking financial sustainability. BRAC is seeking to overcome this exclusion via its 'target within a target' Income Generation for the Vulnerable Group Development (IGVGD) programme, and via enterprise loans for poverty graduates (see the papers by Rahman and Wood in Part IV for further commentary on whether lending to the poor can generate employment for the less entrepreneurial poor).

The paper by Rashid on ASA's 'Self-Reliant Development Model', though shorter and perhaps less self-critical than the others, reveals a deliberate concern for compatibility between the sustainability of lender and borrower. It traces its own evolution since its formation in 1978, revealing a tendency to 'travel lightly' by being prepared to discard earlier strategic stances of social action and legal aid in favour of its current self-image as a 'managerial dynamo' with a lean credit delivery model emphasising: decentralised decision-making; standardised operations; discipline; and the efficient use of funds. This evolution and re-imaging is in contrast to organisations like BRAC or Proshika which have sought with more or less credibility to retain original philosophy as other programme features were added to their respective portfolios. The paper describes its scale of operations and its strategy for achieving financial self-sufficiency among its decentralised units (or branches) via revolving funds with only minimal capital injection from the central office. This model has become very attractive to donors with an agenda of NGO conversion to MFI. The paper then gives two examples of borrower sustainability (paddy husking and puffed rice) over a five year cycle, though no reference is made to the broader market context and the dangers of hidden self-exploitation inherent in such activity (see especially papers by Rahman and Wood in Part IV). The paper finally concludes with a brief review of three sustainability concepts (institutional, behavioural, and policy), seeing evidence for its own success in replication to other organisations both within Bangladesh and outside the country.

Khan's analysis of the role of public sector banks starts with a clear contrast between the banks' contributions to increasing farm production and negligible contribution to direct poverty alleviation, whereas the reverse is true for NGOs. Given the employment generation effects of increasing productivity in agriculture, it could therefore be the case that public sector banks are contributing in ways more favoured by the kind of analysis offered by Prof. Sobhan in Part I. The paper nevertheless proceeds to describe the poverty alleviation programmes which are supported by the public sector banks, as well as those implemented directly by the banks. In its conclusions the paper explores options for linking banks to NGOs in poverty alleviation lending either via bulk lending to NGOs to pass on to grassroots beneficiaries (e.g. Agrani Bank and ASA) or via direct lending to beneficiaries, supported by NGO intervention at the 'social preparation stage' for a service charge. Khan strongly favours the second option as laying the foundation for a long term direct relationship between the poor and the formal banking sector, whereas the first option risks turning the financing bank into a sleeping partner, further underdeveloping their capacity to engage in poverty alleviation programmes. (See Sharif's paper in Part I for a critical discussion of both of these options.) Khan then comments on a series of management issues before concluding with a strong plea for institutional development support to the banking sector (as has been enjoyed by the NGOs), citing the examples of the SIDA/NORAD funded project supporting the Productive Employment Project of the Agrani Bank (see Wood 1994 'Sirs and Sahibs' for an early review of this initiative) and the ADB-assisted Rural Poor Co-operative Project of the Sonali Bank. For Khan, both of these projects refute the common notion that formal banks are not capable of handling poverty alleviation programmes efficiently. Although this paper is strongly preoccupied with the circumstances of the lender, it is more oriented towards a comparison of modes of credit delivery (involving group versus individual lending, repayment schedules, personnel incentives and enforcement) and the place of banks in the livelihood strategies of borrowers, rather than lender sustainability *per se*. The paper was requested by the convenors of the workshop for this reason and is placed in this volume as an important corrective to the NGO and Grameen Bank monopoly on the poverty and finance discourse in Bangladesh.

Problems of Reaching the Poorest

The third part of the volume focuses upon the problems of the reaching the poorest, a persistent theme of the overview papers and raised again in some of the case study chapters. The two papers here are based upon critical internal analysis by staff in BRAC and the Grameen Bank respectively. Hassan Zaman is concerned with both accurate targeting and participation. His analysis reinforces the findings of others concerning the heterogeneity of the poor. The first part of his analysis shows how a mixture of demand and supply side factors leads to the inclusion of a small group of 'non-target' households in poverty focused credit programmes, and describes the characteristics of properly targeted versus ineligible members in order to set up the later analysis of variations in participation (in credit activities) in terms of these contrasting characteristics. The variables which appear to be significant from his multivariate exercise on 'participation depth' are sex and occupation of household head, presence of other credit delivering agencies, electrification and length of membership. There follows a more detailed discussion of the interaction between these variables and BRAC's intervention strategies, before considering the options facing an agency about non-target members. A strategy of gradual exclusion on the grounds of prioritising scarce resources to the poorest is contrasted to a policy of deliberate retention to support the financial sustainability of the organisation delivering credit due to their superior loan absorptive capacity (a finding consistent with that of Hulme and Mosley). A further reason for retention, less concerned with lender sustainability, is that these more entrepreneurial borrowers may create, through their enterprises, employment opportunities precisely for the 'hard to reach' category with less participation depth (again see papers by Rahman and Wood on this theme). Zaman concludes by offering more suggestions for financial service flexibility in order to meet the needs of the poorest more systematically in a way which addresses their poverty correlates (to refer back to Greeley's formulation). At the same time, he cautions against over enthusiasm for the credit-plus approach with its higher costs of delivery and advises that we should weigh carefully the balance of advantage between credit plus and employment leverage strategies arising from supporting the marginal, but entrepreneurial poor.

Hashemi addresses similar concerns in his note on 'Those Left Behind', reflecting on the shared problems of Grameen Bank and NGOs with large scale micro-credit programmes such as BRAC, Proshika and ASA, observing that between them they provide credit to a quarter of all rural households. He focuses upon the problem of member selection among organisations which include within their objectives: financial viability, high repayment rates, increased membership and group formation. The implication of Hashemi's observations is that the group mechanism would appear to be a strength and weakness of these micro credit programmes: overcoming information and enforcement costs on the one hand; screening out high risk borrowers on the other. He points out that where NGOs like BRAC used to have a lengthy period of mobilisation and 'conscientisation' prior to extending credit in order to ensure a strong sense of solidarity among the poor and reduce the screening out of the weakest, there now seems to be a much shorter period between group formation and the first loan disbursement. This is a key issue which goes to the heart of a donor desire to get 'development on the cheap' by favouring the MFI conversion and a continuous questioning of the value of social development in the absence of 'clear quantifiable indicators of impact'. Multi-sectoral development NGOs in Bangladesh and elsewhere are complaining about this pressure, which would appear to be driving them further away from addressing the needs of the poorest precisely when donors present that criticism as well. Returning to Hashemi's own text, he also refers to the problem of the hard core poor self-selecting themselves out of micro credit programmes as being creditworthy, based on a survey of four villages where Grameen Bank and BRAC were active. The reasons for negative self-selection are presented in the paper. Hashemi surmises that these reasons do interact with 'messages' from credit organisations about the necessity of regular and prompt repayment, and the risks to other group members of defaulting participants. Even though the weekly repayment formula was established to make repayment easy for the poorest, this clearly remains a substantial burden, especially if the loan is project related. Hashemi concludes by reviewing two strategies for 'bringing the poor in': policies which increase local level economic activity (a theme favoured by Sobhan, and those betting on the entrepreneurial poor for employment generation); and those which specifically target the destitute, such as BRAC's IGVGD. Of course we might observe that the problem with the latter strategy is how it fits into the lender sustainability objective, raising again the ethics of deploying a subsidy-dependence index to discriminate against welfarist programmes.

Limitations of Scale

The papers in Part IV focus in different ways upon the problem of scale of micro-credit lending, but in the sense of size of loan rather than aggregate numbers of credit recipients. Matin's 'Renegotiation of Joint Liability' is mainly a technical note focusing upon the enthusiasm of the imperfect information school for group credit arrangements to reduce transactions costs for lending institutions, and identifying the conditions under which such arrangements break down. He points out that joint liability is a contract in which the private good of individual access to credit is made conditional on the provision of the public good, namely group repayment. Not only is it supposed to overcome information constraints through peer screening of bad risk, but it also performs essential enforcement functions too (see Hedrick-Wong et al above). He critiques the arguments that such peer monitoring leads to welfare gains for members by reducing risk and therefore interest rates of lenders by pointing out that peer monitoring itself is not a cost-free exercise and that mutual enforcement within a community is not unproblematic. He then makes several observations: with rapid turnover on repayment expected, project based lending demands exceptionally high rates of return just to service loans; that compliance with frequent repayment rates is better understood as an acceptance of credit as advance against future savings made in small, weekly quantities; that peer monitoring rations credit internally against repayment performance rather than the quality of project proposals; than screening is inevitably mediated by other social relationships locally; that conformity with peer pressure can force recalcitrant members into short term, high interest borrowing from local informal credit markets; that peer pressure is more likely a reflection of lender staff pressure which is a more effective sanction than termination threats in a weekly repayment context. He then asks the basic question: if it is staff pressure which triggers peer pressure, is this sustainable? The significance of this question lies in the explanatory problem, for example, of whether the Grameen Bank successes on repayment are achieved by the 'voluntaristic' behaviour of joint liability group behaviour or by large numbers of diligent staff, motivated by performance incentives. Any optimism about the replication of this model (e.g. as a result of the MCS?) relies heavily on which answer is given to this question. It also raises questions about whether joint liability can therefore cope with the strategy of an ever expanding virtuous circle of increased savings capacity (i.e. a form of poverty graduation) as the route to borrower sustainability. In short, is the model consistent with larger scale enterprise which might lift the poor out of their ghettos of self-exploitative employment which the smaller scale loans appear to induce?

These questions are raised by both Rahman and Wood who make pleas for scales of lending which acknowledge the heterogeneity of the poor and in particular the potential for the entrepreneurial poor to generate employment for those hopelessly placed in product, labour and credit markets. Rahman focuses upon the central NGO claim that credit for micro enterprise is breaking the vicious circle of 'low income, low saving, low investment, low income'. As she states: 'The profitability of micro enterprises provides a crucial link between poverty and the extent of access to NGO credit.' She repeats the finding that NGOs are failing to reach the poorest, and reviews the prevailing hypotheses to this effect. She attributes exclusion (including self-selected exclusion) more to demand side factors and endorses the Hulme and Mosley finding that the scope for profitable use of micro-credit may vary depending on the initial situation of poverty and that this will determine the demand for credit. Like Zaman, she is concerned to distinguish between the poor more on the basis of their endowments than entrepreneurial propensity, especially the easy availability of family labour, but also other assets such as homestead, house and livestock. She then refers to case studies to illustrate her points about the significance of the household level, human capital variable (gender, skills, health, and so on). She maintains that borrower graduation is in effect constrained by the NGO/supply side preference for self employment, which is reinforced by the small sizes of loans so that the marginal product of most non-agricultural activities taken up by the poor is so low that the use of hired labour at the market rate would not be profitable (thus implying, of course, an element of self exploitation in family based, small scale enterprises). This experience of small scale lending among NGOs thereby inhibits them from larger loan sizes. However, Rahman is concerned to point out that larger loan sizes may serve the interests of both borrower and lender sustainability by enabling the entrepreneurial poor to employ those poor who are less able to enter product and trading markets (interlocked against them), while simultaneously delivering economies of scale in loan administration thereby reducing lender transactions costs.

Wood's 'Breaking out of the Ghetto' pursues similar themes. He makes similar points about differentiating the poor and critiques the assumption that all poor people can be entrepreneurial within the micro enterprise, micro-credit panacea. This assumption confines credit to non-structural outcomes in terms of a more fundamental transformation of the political economy. This position is shared with Sobhan, and in effect laments the passing of a more radical agenda to address poverty removal in contrast to poverty alleviation. It accepts that it is rational for some poor people to be more risk averse than others, and that markets are highly imperfect so that other factors such as kin networking may be a larger determinant of success within them. The paper calls on the credit community to make a virtue out of this differentiation among the poor to encourage a more flexible flow of resources within mobilised groups (internal on-lending, internal transfers, pooling and so on) to enable the more entrepreneurial to generate employment for others. As for Rahman, this clearly entails larger loan sizes to facilitate entry into more capital based activity. The corollary of this argument is a more differentiated financial product from suppliers to match the poverty correlates (to use Greeley's term, again) of the differentiated poor. Wood uses this position to make a sharp, polemical contrast between 'ghetto' credit (with a high, self-exploitation element as margins are driven down in locally overcrowded, petty markets) and 'structurally significant' credit (or other financial services) in order to draw attention to limitations of micro credit in pursuit of graduation for the poor, though the former may be validated in terms of immediate welfare functions. He then proceeds to critique the over-reliance on credit as a vehicle for sustainable graduation via economic activity, pointing out the features of the political economy which mitigate against the weakly resource-profiled poor, and the need for social mobilisation to strengthen such resource profiles rather than restricting social mobilisation to servicing the group management of ghetto credit. The final two sections of the paper then point out that: first, any prognosis about micro enterprise and employment generation will at least partially be determined by macro-economic performance; and second, there are lessons from other experience (e.g. AKRSP in Northern Pakistan) about relaxing the principle that all group members should gain equal access to, or utility of, institutional loan entitlements in order to match more accurately individual resource profiles (endowments and capabilities). Since the paper refers to 'resource profiles' and room for manoeuvre in institutional employment options, two annexes are provided to explain these concepts.

Micro Credit: A Restricted Approach to Financial Services

The final part of the volume addresses the limitations of micro credit in relation to a range of other financial services and modes of their provision. Within the two papers (Wright, Hossain and Rutherford; and Rutherford), there is a strong emphasis upon the role of savings. Wright et al focus explicitly on the case for voluntary savings as contributing to both borrower sustainability and, ultimately, lender viability. The paper reviews experience of providing flexible financial services to the poor both in Bangladesh and abroad. The point of departure is that the large NGO/MCIs in Bangladesh are using compulsory savings as a source of capital for loans (indeed functioning explicitly as collateral for such lending, as is also the case for AKRSP in Pakistan), while there is pressure from members for access to their savings. Conventional wisdom, such as the donor pressure on Proshika to reduce member access to savings in order to develop its own capital base for lending, represents 'open access to savings' as a dilemma of interests between borrower needs and the sustainability of the lender, with the balance of argument favouring the latter in order to provide any kind of service at all, i.e. without control over savings, financial institutions are presumed to have no incentive to enter credit markets. Wright et al argue to the contrary that open access and other flexible savings facilities may well increase the net savings deposited. The authors describe the case of BURO, Tangail which has implemented a programme emphasising savings instead of credit and provides its members with open access to savings. Their study set out to examine what contribution these savings facilities make to the provision of key financial services to the poor as well as to capitalising the organisation's activities. Their findings, presented in the paper, are that voluntary, open access savings can raise funds similar to those MFIs operating compulsory, restricted access schemes, and may even have the capacity to exceed them. At the same time, the authors note that BURO, Tangail itself has a long way to go in liberalising its services and achieving an acceptable ratio between in-flows of capital from voluntary, open access savings, and the costs of their administration. The paper concludes with a strong plea to the larger MCIs to provide a wider range of financial services in order to match the poverty correlates of the poor and as a route to the indigenous capitalisation of their systems.

The second paper in this final part, by Rutherford alone, naturally shares the policy perspective above but adds a crucial dimension through analysis of 'indigenous' experimentation with mutual financial services. His paper describes the activities and performance of informal finance '*samities*' (associations) in some Dhaka slums, and asks what these informal schemes can tell us about the demand for financial services among the urban poor before speculating on the implications of this experience for NGOs. He distinguishes between ROSCAs (rotating savings and credit associations) and 'fund *samities*' (which are sometimes referred to by others as ASCAs: accumulating savings and credit associations). ROSCAs, as is well known, rotate savings among members as soon as they are created, whereas 'fund *samities*' accumulate savings and lend to members on demand. ROSCAs are generally regarded as safer, with lower transactions and compliance costs among the members, while 'funds' may be both more flexible but also run the higher risk of muddle and fraud. Rutherford then draws on his comparative knowledge of similar schemes in other South Asia cities to predict periods of failure and learning as associations get established and variant practices are tested, followed by trends either towards more professional management and/or toward more city-wide standardisation. A strong conclusion is that these 'poorest' among the poor have an overwhelming need to manage fluctuating cash resources on a day to day basis in order to meet basic livelihoods requirements, and that NGOs should see this as a criticism of the relevance of their emphasis on productive loans. He argues that financial services are essentially about creating lump sums (liquidity) out of normal income flows, and that whereas loans create lump sums now in return for forgoing some of the borrower's future income, savings create liquidity by the saver forgoing some income now for a lump sum later on. While a good financial services programme will enable customers to use both ways of creating liquidity, at present the large NGOs and Grameen Bank only offer the loan option, whereas funds and ROSCAs offer both. This flexibility acknowledges that the poor do have a propensity to make provision for themselves when they can via savings, but they will only do so when they can be certain that they have instant access to cope with day to day fluctuations in incomes and needs. He finishes with some speculation about the future relationship between these different 'provider' institutions.

The concluding chapter to the volume is based upon a re-working of the summary prepared by the editors (with the support of the steering committee) on the third ('rest') day of the workshop for presentation to a larger invited audience on the final, fourth day. The conclusion maintains an initial focus upon borrower sustainability and seeks to distinguish clearly between three strategic options: credit alone; credit plus; and credit with social development. There is an overall preference, certainly by the editors and IDPAA but shared by many of the authors and workshop discussants, for combining a strategy of wider, flexible financial services with a recognition that the value of such financial services can only be secured in a sustainable way by wider forms of intervention in the political economy via various strategies of social mobilisation and conducive macro-economic management. However, while this might be an appropriate perspective on borrower sustainability, there remains the hard-nosed question about the conditions for lender sustainability especially under circumstances of increasing coverage and the need for improved access to sources of capital. The conclusion indicates the policy reforms required to secure that access in the pursuit of poverty removal. The challenge ahead might be posed in the following question: can progressive agencies 'do' social development and be sustainable through the management of other people's money at the same time? As the title of the volume asks: who needs credit?

PART I
OVERVIEW

Chapter 1

POVERTY AND FINANCE IN BANGLADESH: A NEW POLICY AGENDA

Iffath Sharif

Introduction

Credit has always been used as a key element in the development strategy of Bangladesh. Since the early 70s, targeted and subsidized rural credit programs were dominant state interventions to help with the development process. In the mid 70s though, neo-classical economists identified these programs as failures. They pointed out that subsidized interest rates led to an excess demand for credit, followed by the rationing of credit, thus causing distortions of the market (Shaw, 1973; McKinnon, 1973). Further, access to subsidized credit was skewed to the wealthier rural population, thereby leading to less optimal allocation of scarce investment funds. This resulted in a process of financial repression - interest rates on savings deposits were depressed, inhibiting mobilization of domestic resources—which was injurious to the country's overall economic development (McGregor, 1988).

During the same period a number of non-governmental organizations (NGOs) and the Grameen Bank pioneered alternative credit delivery mechanisms for the rural poor that consisted of small amounts of collateral-free, affordable loans, popularly known as micro credit. These micro-credit programs, unlike previous state efforts, have been successful in providing commendable access to credit by landless women and men (usually defined as those with a land-holding of less than half an acre), and in achieving high repayment rates of up to 98 percent. Impact studies on micro credit programs have substantiated the important role of credit in the development process of Bangladesh. The recent BIDS study on Rural Poverty (1996) notes the higher rate of growth in per capita income observed for micro-credit recipients when compared to that of non-recipients. An increase in self-employment, generated by the poor with greater access to credit, has also increased rural wages due to a reduction in the rural labor supply, as reported by a World Bank study on the impact of the Grameen Bank (1995).

The popularity of micro-credit for the poor as a part of the overall development process of a country has spread world-wide. There is almost a global consensus on the importance of micro-credit services for the poor. Most bi-lateral and multi-lateral donors are keen on funding the micro-credit projects of both governmental and non-governmental organizations. With the popularity of micro-credit having reached global proportions, any policy changes regarding credit for the poor inevitably has an impact on any remote micro-credit institution (MCI) in any developing country. This paper will look at one such policy that has received global attention and gradual acceptance but remains to be analyzed in the context of Bangladesh.

The 'New World of Micro Finance Evangelism'

A new vision for MCIs has emerged that encourages MCIs to transform from foreign grant-dependent organizations to large profitable providers of banking services to the poor. The conviction of these micro finance 'evangelists' (Rogaly, 1996) is that there exists an enormous demand for micro-credit by the billions of the poor which can only be met on a commercial basis. Thus the two key emphases of this vision are the achievement of financial sustainability of the MCIs and of maximum outreach[1]. Further, these 'evangelists' make the case that MCIs will best serve their growing portfolios if they have access to formal financial markets (Pischke, 1995; Schimdt and Zeitinger, 1995; Christen et al, 1994). New technologies for improving the financial viability of MCIs are, therefore, being developed and

---

[1] Outreach is measured by the number of borrowers covered by the credit funds (Pischke, 1995).

adopted globally. A growing number of MCIs around the world have started to engage in full-cost pricing of their services, aiming to develop sound financial positions. Subsequently, MCIs are being encouraged to follow a "graduation" path into the formal financial sector, whereby they can diversify the range of their financial services and take on the more sophisticated identity of Micro Finance Institutions (MFIs). There are however, three questions connected to this new vision, which should be raised when assessing its implications on MCIs in Bangladesh. One, to what extent does this new vision enable MCIs to fulfill their role as poverty-reducing agents? Two, do the existing MCIs have the capacity to follow this "graduation" path into the formal financial sector? Three, is the policy environment in the formal financial sector conducive to such graduation?

The emphasis on sustainable credit or other financial services is justified on the basis of the need to strengthen institutions and subsequently increase their outreach, but the resultant borrower situation is equally important. The question to ask is whether the above financial strengthening process laid out for MCIs also brings about the graduation of the poor borrowers out of poverty? Micro finance 'evangelists' believe that access to credit by the poor helps to break the poverty cycle through increased investment and income generation. Recent research, however, indicates that credit for income-generation does not bring the expected results for all poor borrowers. The wider economic environment, the initial social and economic positions of borrowers and their respective debt capacity result in differential effects on their ability to generate increasing and sustained levels of income. Further, the poor are subject to routine vulnerabilities that cause income erosions, and consequently prevent their "graduation" out of poverty (Rahman, 1996). Consequently, maintaining a sound financial relationship with the poor, who face constantly changing poverty levels, can be a challenging task. Factors such as borrower characteristics, living conditions and debt capacity have to be accounted for when developing sustainable micro credit and other financial services directed towards poverty reduction.

Financial strengthening of MCIs to provide financial services to the poor, however, requires appropriate institutional development and a conducive policy environment. MCIs need training and their capacity building in financial management, cost-reducing innovations and accounting practices in order to transform themselves into MFIs. The formal financial sector in Bangladesh, on the other hand, is underdeveloped, non-competitive, and oligopolistic (World Bank, 1994). The quality of existing financial intermediation is low. The sector is dominated by nationalized commercial banks (NCBs) that suffer from poor loan portfolio quality. For MCIs to graduate into a formal financial sector given the prevailing conditions, government efforts are required to increase competition and impose appropriate regulation in this sector through fundamental policy and legislative reform.

The contention of this paper is two-fold. One, that the new world of micro finance 'evangelism' should not only encourage financial strengthening of MCIs to increase outreach, but equally emphasize the need for developing appropriate services that result in the graduation of borrowers out of poverty. Two, the overall financial sector situation in Bangladesh is not conducive to the agenda of MCI graduation into MFIs.

Next, in section two, the paper uses a holistic approach to understanding poverty in Bangladesh which conceptualizes the vulnerabilities and income erosion problems encountered by the poor. The purpose is to understand borrower issues and their variegated needs for financial services. Section three analyzes the implications of the new world of micro finance evangelism given the prevailing poverty, institutional and policy context in Bangladesh. Section four explores ways to complement this agenda to make it a more effective and realistic poverty-reducing policy in the context of Bangladesh. The paper concludes that there is a need to engage in further innovation and institutional experimentation to devise financial services that strengthen the abilities of poor borrowers to graduate out of poverty, and that there is also a need for appropriate policies and regulations by the government of Bangladesh to create a conducive environment to allow MCIs to graduate into the formal financial sector and be able to provide these financial services.

The Poverty Setting

The most commonly quoted measure of poverty is the number of households below a poverty line (the "head count ratio"). This poverty line is calculated on the basis of the level of expenditure required to attain a minimum calorific requirement. The Foster Greer Thorbecke (FGT) class of poverty measures allows one to capture the incidence of poverty ("the head count ratio"); its depth (the average shortfall in expenditure below the poverty line); and its severity (reflecting income inequalities amongst the poor) (Ravaillion, 1992). However there are those who believe the "expenditure" method encompasses only a limited domain of the poverty spectrum. While its prime advantage is the relative ease of measurement, a limitation of the conventional poverty measure is the inadequate cognizance given to non-income dimensions of poverty. Poverty is a multi-faceted phenomenon determined by both material and non-material well-being. Such a holistic approach to understanding poverty is important as it has significant implications for identifying effective strategies for poverty reduction.

Nature of Poverty

According to Sen (1992), to assess the well-being of a person, or to see to what extent she is poor, we need to look at the quality of her life. This assessment can be done by looking at the person's "functionings." Functionings stem from situations such as fulfilling one's basic needs of good nutrition and good health. Higher levels of func tionings include being happy or having self-respect. A person's state of living constitutes interrelated functionings. For instance, having an adequate food intake may not necessarily mean that a person is in good health. That person may not have the knowledge to prepare the food nutritiously. It is only when the person has access to food and is able to prepare it well that we can consider her to be nutritiously fed.

By looking at a person's functionings, therefore, we can evaluate her quality of life or her poverty situation. The various combinations of functionings available determine her capability to lead one life over another. Capability is a person's potential to lead a certain life, while her functionings are the actual occurrences that can be aggregated to assess to what extent that person is able to lead that life. For instance, we can take the example of two capability sets C_1 and C_2, where C is the set of all possible functionings in life. Thus we have the following:

$$C\{f_1, f_2, f_3, \ldots, f_n\}$$

where f_1 to f_n represent functionings of being adequately nourished to being an influential personality in one's community. Let us suppose C_1 and C_2 are comprised of the following functionings,

$$C_1\{f_1, f_2, f_3\} \text{ and } C_2\{f_1, f_2, f_3, f_4, f_5\}$$

where

f_1	– adequate nourishment	f_5	– education
f_2	– good health	f_6	– longevity
f_3	– housing	f_7	– self-confidence
f_4	– access to clean water & sanitation	f_8	– social status

Person 1, whose life comprises C_1, is potentially poorer than person 2, whose life faces the capability set of C_2. Person 2 has greater freedom to pursue access to clean drinking water and sanitation, housing and education, to upgrade her living conditions, unlike person 1, who does not have access to these things. Person 2, however, does not have the chance to pursue longevity, self-confidence or social status. While C_1 and C_2 delineate varying potentials of two persons to lead their lives, whether each is able to achieve their potential is yet another issue. Determinants of poverty are comprised of one's existing potential to lead a certain life as well as one's actual ability to lead it. Poverty is determined by the constraining factors in a person's opportunity frontier. Poverty reduction, consequently, is about being able to achieve the full potential of one's life, as well as improve one's freedom to achieve additional combinations of various elements of living, or functionings, that constitute one's desired well-being[2].

Using the capability set approach to understand poverty is advantageous in that it helps us to move away from a commodity space onto a space of constitutive elements of living. Such a space allows us to identify the various deprivations and vulnerabilities that define the poverty experience. For example, Chambers (1994) identifies several living conditions with which the poor identify themselves that cannot be captured in a commodity space. These include social inferiority, isolation, ill-health and physical weakness, powerlessness, and humiliation.[3] Explaining poverty within a commodity space is limiting as it can not accommodate the multi-dimensionality of the above deprivations and vulnerabilities. Using Sen's capability approach, these deprivations can be looked at as negative function-

---

[2] According to Sen, freedom of choice is important. Choosing a lifestyle is not the same as leading one no matter how it was chosen. One's well-bring depends on how his/her lifestyle emerges as a process of their ability to choose.

[3] The definitions of these deprivations are given in Chambers, 1994, p.13-14.

ings that result from the inability to use or avail oneself of the means to transform these functionings into positive ones.

Who are the Poor?

The above analytical framework helps to explain the multitude of poverty determinants in Bangladesh, and thereby substantiates the diverse experiences in the lives of the poor. Rahman (1995), based on his work on rural poverty in Bangladesh, acknowledges that the poor are not a homogenous group but three major sub-groups can be distinguished. He refers to "extreme poor" households, living 40 percent below the poverty line income at Tk. 3,757 (poverty line estimated at approximately Tk. 6,287 per person in 1994), "moderate poor" households, living just below the poverty line income; and the vulnerable non-poor households, living above but close to the poverty line and constantly facing the risk of falling below the poverty line. Rahman uses a commodity space to assess diversity among the rural poor. The criteria he uses to determine the varying degrees of poverty include income levels, land ownership, housing and self-evaluated deficit status of households[4]. Nevertheless, the three groups of poor that Rahman discusses—"extreme poor", "moderate poor" and "vulnerable non-poor"—are representative of three broad bands of capability sets—lower poor, middle poor and upper poor respectively.

The capability sets for the lower poor (C_{LP}) comprise minimum or below basic survival conditions. These capability sets may constitute zero positive functionings. The poor, at this stage, would strive to generate income to purchase the basic necessities: food, clothing, and shelter. In the capability sets of the middle poor (C_{MP}), they may have the potential to fulfill basic survival needs functionings. The capability sets would incorporate interrelated functionings of food, education, health and so forth. The next desired functioning for the middle poor leading a life constituted by C_{MP} would be that of security. At this stage, the poor want to acquire assets to diversify their means of meeting their subsistence needs. For those whose capability sets constitute both the basic survival and security functionings, their goal will be to achieve increased income generation or the

---

[4] This indicator helps to identify chronic deficit status which indicates extreme poverty, occasional deficit status which indicates moderate poverty, break-even and surplus status which indicate being at or above poverty line respectively.

means to achieve higher levels of functionings such as respectable social status. These capability sets characterize the lives of the upper poor (C_{UP}). Effective poverty reduction strategies need to keep these three levels of poverty in mind to develop appropriate mechanisms to cater to the respective needs of the three groups of the poor, with the over-riding goal of leading people out of their respective poverty levels until they "graduate out of poverty." The size of each group can be an important measure when assessing the impact of poverty reduction strategies on the overall poverty situation. According to BIDS's update (1996) on poverty trends in Bangladesh, in 1994, the extreme poor constituted 22.5 percent of the rural households, moderate poor 52 percent, and non-poor households constituted 26.5 percent of the rural population.

The Economic Graduation Process of the Poor

The economic graduation of the poor can therefore be thought of as the upward movement of the extreme and moderate poor along the poverty-pyramid into non-poor status, and the ability of the vulnerable non-poor to sustain their position. Rahman and Hossain (1995) explain the economic graduation process by making the distinction between reversible incremental changes in household income that result in cyclical mobility within the broad parameters of poverty, and sustained changes in income which lead to a definite rise out of poverty.

There are, however, many obstacles to the economic graduation process that cause cyclical mobility within the poverty line. Rahman identifies these obstacles as "downward mobility pressures" arising from various vulnerabilities and deprivations. The first downward mobility pressure arises from structural factors within the rural economy. The slow growth of agricultural output in Bangladesh, slower than the demand for goods and services produced by poor producers, poses a constraint to production and returns (Osmani, 1989). As a result, some producers are bound to find themselves in situations where the rate of return is less than the cost of borrowing. Demand and supply constraints are also rampant within the structure of the rural economy of Bangladesh (Sharif, 1994). Such structural constraints put a downward pressure on income. Further, seasonal problems of low demand for labor, products and services are not reduced by the provision of financial services. Changes in the wider economy, thus, are crucial for the success of credit-induced activities. The second downward mobility pressure which has particularly been emphasized in Rahman's work on poverty in Bangladesh, is the crisis factors that lead to income erosion problems. Some of the crisis situations include sickness-related expenditures, natural disasters like floods or droughts, theft of assets and other contingencies. Life cycle factors such as the increase in the number of dependants in a household, or the death of an earning member give rise to the third downward mobility pressure (Rahman, 1992; Hulme and Mosley, 1995). These downward mobility pressures may also be mutually reinforcing, thus preventing the poor from economic graduation.

Poverty, therefore, is not only about having inadequate income or income below the "poverty line", but is also the inability to sustain a specified level of well-being. Effective poverty reduction strategies should therefore provide the means or the instruments to achieve one's capability or ability to be well-nourished, healthy, educated, productive, respected and so forth. Once these functionings are achieved, one can improve one's well-being by injecting additional positive functionings into one's capability set, and subsequently move to a higher capability set or a higher state of well-being. These means to achieve one's well-being, although assessed in a space of constitutive elements, can be sought in a commodity space. For example, micro-credit is identified as one such means to help the objective of improving one's well-being by helping to access additional functionings that result in access to education. The point to stress here is that to secure the successful achievement of additional functionings, the existing functionings in one capability set should be fully achieved. For example, before providing credit with the intention of better income levels, one needs to provide the right education about the proper utilization of credit which will result in an improved and sustained income level. Micro-credit strategies need not only emphasize incremental improvements in income levels, but equally need to pay attention to sustainable income gains.

Two predominant poverty reduction strategies followed by most MCIs are social intermediation and financial intermediation. Social intermediation involves organizing groups of poor people in order to offer services such as human and skill development training, health, safe-drinking water and sanitation etc. Financial intermediation, on the other hand, primarily constitutes group formation for savings mobilization, and the subsequent provision of credit. Social intermediation services are more likely to offer direct additions of extra functionings while financial intermediation services allow means to achieve desired functionings. Most institutions follow a combination of these two strategies, with an emphasis on one or the other. This paper will not attempt to discuss the comparative benefits of either strategy, nor what is the optimal level of mix between the two. Existing research work does not give any definite conclusion on this subject. But what one can at least suggest given the poverty context in Bangladesh is that the goal of any poverty reducing strategy should be to prevent the reversibility of incremental changes in the well-being of the different categories of the poor households, thereby allowing for the economic graduation of the poor. Unless these strategies prevent income erosion due to a host of factors, it is highly possible that some of the poor will be unable to improve their living conditions and will fall victim to cyclic mobility within the broad parameters of poverty. It is within this poverty context that the paper analyzes the implications of the policy agenda of the new world of micro finance 'evangelism' on MCIs in Bangladesh.

Whose Graduation: Borrowers' or Institutions'?

The new world of micro finance 'evangelism' encourages MCIs not only to be effective channels of credit disbursement to the poor, but also to become sound financial institutions to scale up their operations and assure their long-term survival independent of donor assistance. The emphasis is on the need for MCIs to adopt principles that require them to break even or turn a profit in their financial operations, and consequently raise funds from non-subsidized sources like the capital market (Rhyne & Otero, 1994). It is argued that providing sustainable and widespread financial services is the only way to reach the millions of poor who still do not have access to credit. The "graduation" path recommended by the new world of micro finance 'evangelism' to achieve the above goal entails a two-step process. MCIs are encouraged first to fully cover the cost of their operation[5], and eventually graduate into the formal financial sector. It is argued that the sound financial positions of the MCIs will allow them to access the local capital market. Accessing commercial sources of

---

[5] Cost of operation includes salaries and administrative costs, depreciation of fixed assets, the cost of loan and principal lost to default constitute the non-financial costs, while financial costs include the cost of funds used for on-lending (World Bank, 1995).

funds helps to maximize the leverage of donor funds[6] that MCIs may have been receiving. By graduating into the formal financial sector MCIs can fund their loan portfolios fully in commercial financial markets, either by capturing individual savings deposits or by attracting investors through the issuance of debt securities. Such a process will allow a dollar of donor investment to leverage up to eleven to twelve dollars of micro finance assets after a few years. Consequently, MCIs are able to maximize their outreach (Pischke, 1995). The following section addresses three key constraints faced by the above policy agenda of the new world of micro finance 'evangelism.'

The Trade-off between Poverty Reduction and Financial Sustainability

The new world of micro finance 'evangelism' identifies financial sustainability as a prerequisite for the widespread availability of credit services. According to Rylme and Otero (1994) financial sustainability is defined as covering operational costs, including loan loss reserves, the cost of funds and inflation through fees and interest charges. Profits are encouraged for achieving genuine commercial viability. Loan loss reserves are necessary since delinquency and default cannot be eliminated totally, while inflation is required to be built in to interest rates. Costs of funds will include any interest paid on savings generated by the MCI or if loan capital is accessed for credit operation. Cost of funds will be null for MCIs who are dependent on donor grants. Services that help to prepare borrowers to manage and use credit are considered to be a necessary cost of lending to the poor and are therefore built into the costs. The assumption is that borrowers are far more sensitive to the availability of credit than to the interest rate. As the first step of the economic graduation process, MCIs are required to maintain full-cost pricing policies. Can we be sure, however, that improved access to full-priced credit services through the above process will result in the economic graduation of the poor?

Recent experiences of MCIs have shown that focusing mainly on institutional issues brings about a risk of excluding the extreme poor

---

[6] Leverage is defined as the ability of a program to use its capital (donor funds in the case of credit schemes) as a lever to obtain additional funds through borrowing to taking deposits. (World Bank task Force report on "Poverty Alleviation Micro Finance Program," 1995).

from credit programs. Increasingly, the extreme poor are seen to be dropping out of credit programs after having failed to keep up with repayment installments (Hulme and Mosley, 1995). One reason can be attributed to demand side effects. For instance, when the poor receive credit, their returns on investment have to be adequate to meet the repayment installments from the incremental income of the activity undertaken (Rutherford, 1995). This is known as the break-even rate of the poor (Arora & Upendranath, 1995). In the event that borrowers are unable to earn this level of income, they are faced with the following options to finance their repayments: borrow from other sources; use savings or sell assets; reduce existing consumption levels; reschedule the loan repayment if possible; or default on the loan. Households, for whom none of the first four options are feasible, will be forced to drop out of the micro-credit program. For other households, the lack of other sources of income will diminish their confidence to go into debt and they will thus become part of a self-exclusionary process. It is quite likely that borrowers may fall victim to one or more of the above situations, given the "downward mobility pressures" which are part of rural economic reality, as discussed in section two.

Supply side factors, on the other hand, help to exclude the extreme poor from credit programs. Institutions seeking sustainability objectives will be inclined to focus on a wealthier clientele to guarantee full repayment of what will be costlier credit services by virtue of the full-cost pricing policies. Pischke (1995) points out that sustainability targets will cause the focus on borrowers to shift from the poorest of the poor to the less poor. According to him,

> "Emphasis on financial analysis and financial viability could logically lead to a subordination of outreach in order to become profitable. One means of doing this is to abandon the original target group. This generally occurs through subtle migration, not as an abrupt change in course, possibly through an increase in average sub-loan size or through development of new services that appeal to a clientele that is more wealthy than those whom the lender originally undertook to engage." (p. 21).

Thus, institutions trying to achieve greater levels of self-sufficiency through full-cost pricing of credit services may be doing so at the expense of extreme poor borrowers. One cannot help but question the integrity of organizations trying to achieve parallel objectives of poverty reduction as well as their own institutional financial sustainability at the expense of poorer borrowers. With 22.5 percent of the rural population of Bangladesh living under extreme poor conditions, fulfilling the credit needs of only the "wealthier" poor will shift the burden of poverty onto those who are already over-burdened with it. Distribution of resources will worsen, causing no particular dent in the overall poverty situation. Micro-credit and other financial services to the poor not only need to target the sustainability of the institutions, but equally need to guarantee sustainable benefits to all poor borrowers, particularly the extreme poor. *Further research therefore, is required to develop micro-credit and financial services that will address issues of borrower vulnerabilities and thus, help to accomplish borrower graduation out of poverty.*

Structural Obstacles to Institutional Graduation

The second step proposed by the new world of micro finance 'evangelism' involves the graduation of MCIs into MFIs. Micro finance 'evangelists' expect such graduation into the formal financial sector to help expand their resource base by accessing the local capital market (Schmidt and Zeitinger, 1995). MCIs, however, would need extensive restructuring in order to capture deposits and provide credit on a commercial basis. A second structural change needed is the decentralization of the operation for expansion purposes. Thus strong management information systems need to be developed. Other issues for consideration include institutional and staff incentives structure, acceptable accounting practices etc. Finally, legal considerations are required to define how these institutions need to change internally to operate in the formal financial sector.

No MCI in Bangladesh is at present fully self-sustaining and many are far from reaching such status. One major reason is the fact that social intermediation services constitute a major component of MCI programs. Social intermediation is costly, and brings in no returns, unlike financial intermediation. Donor grants or soft loans usually help MCIs to cover their operating expenses and run revolving loan funds. Interest rates and program fees only cover portions of current operations, thus requiring continued donor grant support. The requirement of subsidy is much greater in the case of newly formed MCIs than on-going ones, many of whom have started experiencing economies of scale and are able to cross-subsidize portions of their social intermediation costs with returns from financial intermediation services. The efficiency of these large MCIs has also increased due to computerized management information systems, improved financial management, etc. The second major reason behind low levels of self-sufficiency is the trade-off between outreach and sustainability. Serving larger numbers of people in different regions involves setting up new branch offices which in the short-run cost more to operate than they can generate in interest revenue. Most of the large MCIs expanded rapidly in the early 1990s thereby compromising their short-run viability (Zaman, 1995). However this process is essential if MCIs are to grow into fully-fledged formal financial intermediaries. Thus, MCIs may not be ready for immediate graduation into the formal financial sector. *Intermediate processes need to be emphasized in order for MCIs to slowly achieve incremental levels of self-sufficiency through structural reform as well as experience with working with loan money rather than grants.*

The Formal Financial Sector and the Constraints to MCI Graduation

Cuevas (1995) argues that there exists a set of necessary conditions for MCI graduation. Given his analysis of 'graduation' cases in Latin America, he concludes that successful graduation of MCIs depends on key financial reforms, perceived potential and the actual ability of the institution to enter into a profitable "market niche", the need for appropriate regulatory and supervisory mechanisms and a careful analysis of projects to be funded. Cuevas delineates the following specific terms for financial reforms and deregulation of the formal financial sector:

- market-determined interest rates
- elimination of direct credit allocation
- rationalization of reserve requirements
- reduction of barriers to entry
- modernization of regulatory systems so that the role of government is to monitor and set prudential regulation and supervision.

The government of Bangladesh has pursued the Financial Sector Reform Program since 1989. The World Bank, however, reports that such financial market liberalization, to date, shows mixed progress, and thus prevents proper assessment of its impact on the quality of financial intermediation (1995). The major barriers to the financial sector development in Bangladesh have been the government ownership of financial institutions and state interventions in credit allocation. The quality of loan and recovery rates have over the years been low. Easy refinancing facilities from the Bangladesh Bank, on the other hand, helped to conceal poor repayment performance. Such poor debt recovery is constantly exacerbated by the involvement of vested interests in borrower selection, loan amnesty, and use of credit for relief or patronage purposes. Some of the other reasons for the poor performance of banks include poor management, allocating high risk loans to priority sectors and loan forgiveness programs of the government.

Even though interest rates have been liberalized, they are not fully market-determined for structural reasons. The NCBs are able to execute substantial influence on the market interest rates due to the fact that they enjoy oligopolistic market dominance. The relatively high cost structure of NCBs encourages high lending rates in real terms. The small private sector, on the other hand, enjoys a comfortable position which enables it to earn high margins, making it unwilling to undercut the NCBs. Moreover, as part of the Financial Sector Reform program, the government provides interest rate subsidies to various priority sectors from its own funds and asks NCBs to subsidize "sick" industries. Such erratic interventions by the government, in the name of improving resource allocation, are not conducive to profitable lending. Financial intermediaries need to use commercial criteria for selecting borrowers and projects. Pushing for the graduation of MCIs, therefore, may be premature given the existing conditions of the formal financial sector. *Rather, government efforts should be concentrated on bringing about flexibility in interest rates on deposits and loans, and improved prudential regulations and supervision of formal financial intermediaries.*

The above analysis of the policy agenda of the new world of micro finance 'evangelism' in the context of poverty reduction in Bangladesh brings to the fore three key implications: (i) the need to develop a set of financial services that address not only the financial sustainability of institutions, but also the economic graduation of all poor borrowers, particularly the extreme poor borrowers, out of poverty; (ii) the need for an intermediate process before MCIs are ready for graduation into the formal financial sector, and (iii) the need for an appropriate legal and financial regulatory framework to accommodate the graduation of MCIs into the formal financial sector.

Challenges Ahead

The prescription for MCIs to reach institutional sustainability, and subsequently increase leverage of funds through graduation into the formal financial sector is important for increasing outreach, but not comprehensive in the context of poverty reduction. Micro credit and financial services need to be developed that address the needs of the various groups of poor borrowers. Structural reform at the institutional level and legislative reform at the policy level is required to allow for the successful graduation of MCIs into MFIs. This section illustrates some thoughts on possible options that could address the above concerns.

Flexible Financial Services for the Poor

Section two established the differing conditions of the poor in Bangladesh. The needs of the poor to fight their respective poverty situations should therefore also be varied. Most MCIs, however, are characterized by several limitations and rigidities that reduce their abilities to cater to the needs of the diverse groups of the poor. For example, the poor value the opportunity to make regular savings but have no secure place to save. Thus, whatever limited amounts of savings they manage to put aside tend to get "eaten away", sometimes literally. Having an external and secure place to save is a way of protecting their savings from pressures and demands of everyday family life[7]. Most credit models do not provide fully-fledged savings services to their borrowers. Normally, members are required to save to consolidate their sense of commitment and bonding within the group, and thus be eligible to borrow. Such enforced group savings mechanisms tend to reduce the average level of savings per person because of the barriers associated with withdrawals in times of need[8].

---

[7] The limited opportunities for saving among the poor are shown by the alternate savings mechanisms that the poor have developed. For example, corrugated iron is seen as an investment that can be liquidated in the event of an emergency or sudden cash requirement.

[8] See Zaman *et al* (1994) on BRAC's experience in flexible versus rigid savings mechanisms.

Being able to save individually not only allows the poor to turn to their savings in times of need, but also helps them build their confidence to borrow money. The reason being that weekly repayments are often made out of the regular savings of the poor as repayment schedules are usually not consistent with project income flows. Moreover, given the small amounts of loans, borrowers are forced to engage in subsistence level activities, which then become a supplementary source of income. As a result, borrowers also depend on their main source of income to make weekly repayments. Therefore, it is only those borrowers who have confidence in their ability to make these weekly repayments through their savings or from other sources of income who will be prepared to take up these loans[9].

Small loans are more beneficial to the relatively "better-off" poor, in that they allow the expansion of existing business. These "wealthier poor" are able to invest in stock-in-trade. For them, the return on the additional investment is often enough to repay weekly installments. Small loans thus allow the better-off poor to expand their business while the poorer borrowers may only be able to engage in supplementary livelihood activities (Rutherford, 1995). Greater volumes of credit support could bring in greater returns. Not all the poor, however, have the capacity to utilize large volumes of credit. Support services are required to effectively translate loans into profitable venture. For example, technical services to identify and manage small scale enterprises, information about potential markets, possible marketing channels, and appropriate technology is required. It is only when credit is coupled with the above services that lending institutions can consider increasing loan sizes. Otherwise the risk involved with lending may be greater, though this has to be traded off with the costs of providing these services.

Life in rural Bangladesh is constantly vulnerable to income erosions as a result of contingencies that may be brought about by structural reasons, sickness, death of an earning member of the family, and other unforeseen events. Incidents such as these lead to income erosion problems that may lead to reverse mobility within the three poverty bands. Unless financial services account for these contingencies, borrowers are likely to fall victim to the "drop out" phenomenon, explained in the earlier sections. A possible escape

---

[9] According to Rutherford (1995), a new business would have to generate an internal return of 125% to service a loan with an interest rate of 25% with loan capital to be repaid in one year.

from above situations could lie in providing innovative insurance services against events that lead to income erosions, subject to appropriate regulations[10].

Poor people, therefore, need financial services such as various savings schemes, consumption and investment, credit and insurance services to smooth out their household cash flow, deal with emergencies and other unforeseen requirements of cash and augment income through investment in any gainful way (Ibid, 1995). The challenge, however, is to devise the right group of services for the right group of the poor. For example, extreme poor households are more inclined to be in need of "survival measures" which could include voluntary savings mechanisms and emergency consumption credit facilities. Moderate poor households, on the other hand, need "protectional measures" that offer relatively low-risk income-generation activities and other services such as training and education for the building of their debt capacity. The vulnerable non-poor households constituted by small or medium farmers or entrepreneurs, are in a position to require "promotional measures" which include primarily credit, savings and insurance for income generation, as well as support for product development, technical change, marketing, and support for sharing some of the risks of technical innovation (Hulme and Mosley, 1995). MCIs need to tailor financial intermediation services that account for these survival, protectional and promotional measures.

Collaboration among MCIs, Banks and the Government

Section three established that the policy environment in the formal financial sector in Bangladesh is not conducive for the graduation of MCIs into the sector to access the capital markets. Instead, a collaborative system based on coordination among MCIs, the formal banking sector and the government of Bangladesh could be established to facilitate an intermediate process prior to MCI graduation into the formal financial sector. Such a system will require the cooperation between MCIs and the banking sector, and strong policy imperatives on the part of the government to facilitate that cooperation.

---

[10] For example, Hazell (1992) proposes a simple drought/flood ticket scheme where if the probability of a flood/drought is 10 percent in a given area and the average cost of administration of the scheme is 2 percent of total indemnity value, a ticket could be sold to individuals for Tk. 12 redeemable for Tk. 100 in the event of a flood/drought.

With most MCIs operating at low levels of sustainability, MCIs could be encouraged to operate at one of the intermediate levels before they can be expected to graduate into the formal financial sector. To experiment with their capacity to achieve incremental self-sufficiency, MCIs could start off with raising funds by borrowing from the banking sector, or on easier terms from other sources[11]. Interest income from financial intermediation can help cover the cost of funds and a portion of operating expenses. Grants will still be required to finance some aspects of operations. Some commercial banks like Sonali, Agrani, Janata and the Bank of Small Industries and Commerce (BASIC) are already engaged in financing some MCIs. The two MCI-Bank linkage models that exist are the "wholesale" and the "retail" arrangement.

Under the wholesale arrangement, banks provide loans to MCIs, which are used for further on-lending to the poor. Although the wholesale arrangement is a commercially viable model, MCIs face a major policy bottleneck: collateral; and a cultural bottleneck: "expediting fees" or bribes. To facilitate wholesale linkages between MCIs and other commercial banks, either the Bangladesh Bank or the relevant authorities (the Ministry of Finance) need to do one of the following: (i) encourage banks to waive collateral requirements for MCI loans used for poverty reduction purposes, (ii) act as the guarantors for MCI loans taken for further on-lending for micro credit purposes, or (iii) instruct banks explicitly to allow MCIs to pledge their "borrower-receivables"[12] as collateral based on the MCI's record of performance. Further, since banks may need to familiarize themselves with such a clientele, relevant training is required for the bank staff. MCIs, on the other hand, may have their own accounting practices that are not consistent with that of the banking sector. Necessary training needs to be provided to MCIs so that they are able to provide banks with the required financial information in the required standard format.

In the "retail" model, the MCI is responsible for supervision and monitoring of the loans, while the bank provides loans directly to MCI borrowers. A cut of the interest rate charged is given to the MCI as "supervision fees." One important advantage of the retail model is that it allows poor clients to be in direct contact with the

---

[11] The Palli Karma Sahayak Foundation is a welcome intervention in this respect.

[12] Borrower-receivables refer to the loan outstanding.

bank. Thus, if the MCI pulls out, one would expect the borrower-bank relationship to still exist. This model also allows banks to waive the collateral requirement given the small size of the loans given out to the borrower directly (loan sizes below Tk. 50,000 do not require collateral). However, existing bank procedures for loan applications may discourage or even intimidate poor borrowers (it is costly, requires bribes, lengthy procedures, etc.) as the Swanirvar Bangladesh experience has shown. Besides, overseeing applications for small amounts of loan increases the transaction costs of banks. This may push banks into a loss-making situation, especially when banks have to operate within a lending rate band (9.5%-16%) as per Bangladesh Bank rules. The retail model is infested with other problems. There exist attitudinal problems on the part of bank staff when dealing with poor clients. Further, the terms and conditions placed on the MCI may not always be favorable in that the percentage that they receive may not cover the cost of the MCI's responsibilities of loan supervision, monitoring and recovery. The retail model maybe more appropriate for "MCI graduates", which is small but growing in Bangladesh. This group will want to invest in riskier, higher productivity enterprises, and require larger loans over a longer term with different repayment schedules. Such a need is not fulfilled by most MCIs. Banks could take on the responsibilities of these "MCI-graduates" through a specialized credit scheme. *Not only will this mechanism free-up extra MCI funds to be used for further horizontal expansion, but it also serves the purpose of bringing the poor into the mainstream financial system.*

Given the prevailing institutional constraints in the financial sector, a strong implementing body, like the Ministry of Finance, is needed to exercise its supervisory influence over the banking sector to help accomplish the partnering roles between MCIs and banks. Government policies that will facilitate MCI-bank linkages for poverty lending need to be implemented. This may involve providing policy directives from the Ministry of Finance to commercial banks asking them to waive collateral requirements when lending to MCIs for micro credit lending purposes. Alternatively, the Bangladesh Bank could act as the guarantors of the loans given out to MCIs or request banks to accept "MCI borrower-receivables" in lieu of collateral. The long-term government imperative should be to establish a favorable policy and legislative environment so that formal, semi-formal and informal financial mechanisms are involved in lending for poverty reduction. Such a policy environment will help to extend the reach of the formal financial sector to the poor sections of the population, mobilize new sources of savings and tap new markets for credit, and in the process increase the prospects of sustainable MCI graduation into the formal financial sector.

Conclusion

The policy agenda of the new world of micro finance 'evangelism' encourages MCIs to achieve sustainability in their operations, and eventually graduate into the formal financial system in order to tap into additional commercial funds to increase their outreach. This paper discussed the trade-off between financial sustainability and poverty reduction. The paper argued that issues regarding economic graduation of borrowers need equal emphasis as do financial strengthening of MCIs. The paper also discussed the structural and policy constraints associated with MCIs graduating into the formal financial sector. Most MCIs and the formal financial sector environment do not meet the conditions for the successful graduation of MCIs into the formal sector, and their provision of flexible financial services to different groups of the poor on a large-scale. The paper suggested an intermediate stage whereby MCIs can gain incremental levels of financial sustainability, and eventually graduate into the formal sector. The paper laid out a policy agenda for MCIs, formal financial sector banks and the government of Bangladesh to engage in a collaborative effort that would ensure the necessary conditions for MCI graduation into the formal financial sector. The paper recognizes its limitations in that it is only making observations which are not substantiated by primary research. The observations of the paper, therefore, are suggestive and not conclusive. The significance of the paper lies in initiating debate on the feasibility of the policy agenda of the new world of micro finance 'evangelism' and its implications on poverty reduction strategies in Bangladesh.

Chapter 2

POVERTY AND WELL-BEING: POLICIES FOR POVERTY REDUCTION AND THE ROLE OF CREDIT

Martin Greeley

Overview

This paper suggests three things.

1. That the poverty reduction impact of micro credit programmes be measured by movement above a poverty line based on food consumption. There are many other important things to achieve but they should not be confused with absolute poverty.

2. That in recognizing the success of micro credit programmes in reducing poverty for millions of poor people, the livelihood and welfare needs of many other very poor people needs careful re-assessment. Micro credit, as presently delivered, is not the answer.

3. That, despite apparent convergence, the implementation of the macro-economic reform agenda may undermine the present micro credit strategy.

Introduction

In February 1997 in Washington DC the Micro Credit Summit will take place. The objective is to help energize governments, donors and NGOs to provide credit to 100 million poor households by the year 2005. This substantial, but approachable, target is evidence of the global commitment to micro credit services as a vehicle for poverty reduction. The pre-summit documentation also makes it clear that the empowerment of women is an anticipated corollary of promoting credit services to female customers.

The global enthusiasm for micro credit is undoubtedly influenced by the scale and success of targeted rural credit programmes in Bangladesh. The significance of lessons from the Bangladesh experience has been enhanced by the volume of research that has been undertaken during 20 years of implementing micro credit programmes. Whilst some of this research has been characterized by an unswerving loyalty to the NGO-implemented, targeted joint-liability approach there has also been some very serious analysis of the impact of these programmes and their sustainability.

This paper addresses just two of the issues that this research has highlighted and a third which it has not yet addressed. First, since all programmes claim that they are seeking to reduce poverty, how should poverty reduction be measured? Secondly, as programmes in Bangladesh continue to grow are there any broad lessons about which poor people are not benefiting? Thirdly, how do micro credit programmes relate to the broader policy environment of liberalizing macro-economic reform, private sector development and export-led growth which Bangladesh has of necessity adopted? These three issues obviously provide only a very partial analysis of policies for poverty reduction and the role of credit. The first and the third have been selected because they may not be treated in other papers at this workshop; the inclusion of the second issue, which is covered elsewhere, marks the growing recognition of the importance of understanding differences amongst poor rural households in assessing the benefits from micro credit programmes.

Poverty Measurement

"You must be before you well be"
Michael Lipton

In the 1970s, poverty was usually measured by some form of head count based on an estimate of per capita consumption, expenditure or income in relation to an estimate of minimum necessary consumption. This is essentially a definition of food poverty though, in a less rigorous way, it includes other essential needs (Ravallion and

Bidani 1994). In the 1990s, with the reaffirmation of the centrality of poverty reduction to the development policy agenda, concepts of poverty have become much broader. The idea that poverty is multi-faceted has been widely accepted and, apart from material deprivation, it is not unusual to include health status, illiteracy, and several types of vulnerability, powerlessness and absence of choice. A fundamental ideational change associated with many of these restatements of poverty definition (e.g. CIDA, 1994) is the view that people's own perceptions are fundamental to identifying what poverty is.

This more all-embracing conceptualization has several origins. First, it has empirical roots, emerging from the experience of development projects and from the adoption of basic needs fulfillment as an objective in rural development. Second, it has intellectual roots in critique of the reductionism of economics. The welfare basis of real income comparisons has always been a controversial topic in development economics. The inability to use the metric of utility as a basis for interpersonal comparisons of welfare has left the adherents of the "income equals happiness" school adrift without theoretical foundations. This line of attack is old (see e.g. Little 1957) but more recently is particularly associated with the work of Amartya Sen (e.g. Sen 1982) who has developed a rights-based ethical theory of welfare centred on the notion of capabilities. In this approach, income is of derivative value only as a means to fulfill certain ends; incomes are one of a number of sources of entitlement that allow control over commodities with specific characteristics which in turn enable individuals to achieve certain capabilities and these allow them to realize an individuated set of essential functionings.

Thirdly, as Shaffer's (1996) recent paper describes there are foundational differences, both epistemological and ethical, between the income or consumption based approaches to poverty measurement and newer approaches that are participatory in character. He argues, convincingly, that the naturalist ethics underlying consumption based approaches to poverty measurement are grounded in a different concept of science to the discourse ethics of the participatory school. There is therefore no basis for arbitration when they produce conflicting results (as in Jodha 1988). The popularization of the participatory approach is particularly associated with Robert Chambers (1995) and with innovative field methods to improve the accuracy with which outsiders can learn the realities of poor people's perceptions of need. These constantly evolving methods (Participatory

Rural Appraisal) focus on relations between the internal and the external (emic and etic) both in terms of relationships between the outsider and the "participants" and in trying to distil the realities of poor people's experience of poverty from the socially constructed false consciousness that colours their articulation of well-being.

In view of these developments it may seem rather old fashioned and overly economistic to argue for the need to retain a focus on using food poverty lines to provide consumption based estimates of poverty in evaluating the impact of micro credit programmes. Whilst immediately recognizing that this is not all one should do I believe there are overwhelming arguments in favour of ensuring that impact assessment is grounded in such estimates. There are five substantive reasons.

First, most ethical theories of welfare ultimately recognize some sort of hierarchy of need with material well-being, in the absolute sense of meeting some basic level of physiological need, at the top of the list. This is true of Rawls's (1971) theory of justice, of Doyal and Gough's (1993) theory of human need, of Sen's theory of entitlements (Dreze and Sen, 1989) and of Maslow (1954). Whether in the language of negative freedoms or avoidance of harm, or the outcome of decisions behind the veil of ignorance, the primacy of physiological needs is restated. Personal liberty is the priority in Rawls's account but only after some basic level of material well-being has been achieved. As against this there is no alternative evidence, either from theory or empirical study, to suggest that material well-being is subordinate.

Secondly, this view is frequently confirmed (*pace* Jodha, 1988) by participatory assessments with poor people including assessing impact of micro credit programmes (Mustafa et al, 1996 p. 17) when poor people themselves refer to the material basis of their poverty. In Bangladesh, what has frequently emerged from such discussions is that the length of time that households can feed themselves for over the course of the year is the basis for differentiating between households. Whilst this cannot be directly related to a consumption based poverty line it shares a common basis in prioritizing physiological need.

Thirdly, alternative approaches to defining poverty are often incapable of being aggregated and therefore of providing unambiguous assessment of programme performance. This is obviously true of subjective estimates, whether the aggregation is over space or through time, but it is also true, for example of basic needs assessments. The Human Development Index (HDI) is one example where an attempt has been made to aggregate (literacy, life expectancy and income) but there is no obvious justification for the choice of these particular indicators or that they should have equal weights. Nor does the HDI offer a basis for household level assessment of poverty condition; the UNDP-commissioned HDI for Bangladesh is based on data from greater districts and from *zila*-level and is of little help in prioritizing resource allocation. Attempts to combine two or more indicators together are intuitively appealing because they seem to provide a more realistic concept of poverty but such attempts have so far failed to produce absolute and comparable indicators at household level.

Fourthly, a key characteristic of a poverty indicator is that it should be comparable over space and time. If an indicator is to be used for impact assessment this is an essential characteristic. In effect this means that it must be an absolute measure. (Defining poverty condition relatively, e.g. the lowest 20% based on income, or those with no land, will not allow knowledge of whether there has been any improvement as a result of development interventions.) Consumption in relation to an appropriately defined poverty line fulfils this requirement. In principle, other indicators could also meet this requirement. In practice it is very difficult. For example, it is common in Bangladesh to use some measure of housing quality in assessing poverty condition but as a recent review of different poverty indicators showed (European Commission 1995) there is a bewildering variety of measures used—floor space, building materials, number of rooms etc. Even if definitions were used consistently in ways that allowed comparison—say, floor space was universally adopted—it is quite unclear at what stage one would declare a household to be out of poverty. Similarly, with indicators of empowerment that are comparable through space and time, e.g. numbers of journeys that women make out of the village unaccompanied by a male; determining when they are deemed to be empowered is wholly arbitrary (see Hashemi, Schuler and Riley, 1996 pp. 637-639, for a discussion on constructing empowerment indicators).

Fifthly, sustainability of micro credit programmes is of great topicality and has several components; one of these involves the household-level analysis of material benefits from credit. Programmes will not be sustainable, other than as safety nets, if the net material benefits from participation do not result in reducing food poverty. Ignoring, for the moment, the issue of intra-household food distribution, households in food poverty are unlikely to be running enterprises at levels of remuneration that allow them to accumulate. Even ignoring the risks of non-repayment, and, therefore, the threat of reduced support from efficiency-minded donors, the inability to accumulate and the persistence of food poverty perpetuates vulnerability and is evidence that programmes are not achieving their objectives. Whilst absence of food poverty is only a partial indicator of sustainability, and there are market considerations that also matter a great deal, it remains the case that, where absolute poverty is endemic, this criterion is of paramount significance in assessing sustainability.

A food-based poverty line accommodates all these concerns discussed above. It can be aggregated—indeed it is sometimes referred to as the head count ratio. It is objective, in the weak sense that it is comparable, replicable and universalizable and does not depend on any individual's judgement. Whilst the claim of objectivity may be of little appeal, indeed little meaning, to post-modern thinking, it does suggest a basis for fairness both in resource allocation decisions and in assessing programme performance. It is absolute in that the poverty line is set at a level (roughly 2100 calories per adult male equivalent per day) corresponding to minimum dietary needs (see Ravallion, 1992 and Ravallion and Bidani, 1994 for a discussion of alternative formulations). This is essentially an arbitrary choice—and in other countries where food poverty is no longer a main policy concern would not be appropriate- but in the context of poor countries it does provide a core indicator of progress. It is also the case that, even though utility is not directly commensurate with an absolute poverty line, there is considerable theoretical support in the welfare literature, as discussed above, for the poverty line as a well-grounded instrument of poverty assessment when set at this level.

The reasons that the poverty line has being subject to criticism are partly because of its perceived narrowness but also because, in practice, it has been difficult to provide reliable and timely data to measure poverty in this way. It requires household level information on food intake. Physical estimates would be ideal, as the Bangladesh Bureau of Statistics uses in its Household Expenditure Survey, but often, money estimates of income or expenditure are the source of data. In these cases, additional information is required on shares of food and non-food items in consumption, sources of calories and prices. Poverty lines fail to take account of intra-household inequality in food intakes, which is certainly a weakness in the Bangladesh context, and usually deals poorly with differences due to (adult) age and activity. These are what Sen (1981) refers to as identification problems—as opposed to aggregation problems which characterize many other poverty estimates, notably subjective ones. Whilst they are serious, unlike aggregation problems, they are not insurmountable. There has been considerable attention paid to the improvement of survey techniques including the use of successive rounds of household surveys to produce panel data which confers greater accuracy, e.g. in impact assessment, because the same households are assessed at different points of time. The cost of resources needed to address these problems of survey method are usually small in relation to programme budgets.

The benefits for micro credit programmes from obtaining reliable estimates of food poverty are not restricted to a well-grounded estimate of poverty reduction impact. Using simple alternative estimation procedures (see Greeley 1994) the same data can be used, for example, to establish whether the programme is selectively biased towards the top of the target group (if the head count ratio reduces more than the severity index) or whether the very poorest are getting preference (if the intensity index falls by more than the head count or severity indices).

However, poverty line analysis provides little understanding about the dimensions of poverty, about poverty as a process or any basis for fine-tuning poverty reduction interventions. This understanding comes through the analysis of poverty correlates such as employment status, asset holdings, child nutrition, housing quality, health status and educational attainment; in Bangladesh context, other commonly employed correlates include, female empowerment, access to land, resilience to seasonal stress, vulnerability to natural disasters, personal security, and observance of legal rights. Analysis of these correlates is fundamental to micro credit impact assessment both with respect to particular programme components which may have been expected to have some specific impact and to provide a more rounded view of impact on well being.

Identifying these other indicators of well-being as correlates of poverty rather than as measures of level of poverty is a clear way of separating the question of who the poor are from what are their characteristics. Both elements are essential to a poverty profile but the distinction between them is useful. It is the level of poverty that determines the need for development interventions and establishes priorities between competing claims; it is the correlates that have to inform the design of projects, programmes and policies. Interventions planned solely on the basis of levels of poverty are likely to fail. Resource allocation decisions made on the basis of poverty correlates are likely to be sub-optimal from a welfare perspective.

Which Poor Households do not Benefit?

"Lord, it is proclaimed you have banished poverty from our land I beg to report the villain has taken up abode in our home"
(roughly speaking)... an old Indian verse

The global commitment to micro credit as a vehicle for poverty reduction is evident from the aims of the forthcoming micro credit summit. In Bangladesh, there will be further expansion of these programmes in response to their success in effectively targeting poor people and reducing their poverty. It is believed that programme expansion will strengthen the likelihood that structural constraints of economic, social and gender relations will be reduced (Mustafa et al 1996 p. 20). In other words, that "going to scale" will in itself enhance the prospects for poor households to achieve sustainable livelihoods. However, it is now widely recognized that not all poor households have been able to benefit from micro credit programmes. There are three groups of such households: those who are active members but who are not improving their economic condition; those that initially join but subsequently drop out or become inactive as loan recipients; and, those that never join. This section focuses on two of these groups, dropouts and non-members, because the third category, under-performing active members, are already a major focus in the research literature. These two groups merit special attention because of the obvious risk that the successful models of Bangladesh, and elsewhere, will enjoy an increasing share of the resources available for direct poverty reduction programmes at the cost of those poor households for whom micro credit is less likely to be a part of the solution to their poverty but who may otherwise have benefited from alternative poverty reduction initiatives (Rogaly 1996). Four caveats are needed.

It is important first of all to note that recognizing limitations and risks with micro credit programmes is not a criticism of what they currently do. In fact, precisely the opposite is true. The very real successes of these programmes which have benefited millions of poor households has led to the resources being made available for their expansion. Secondly, there are differences between programmes, both organizationally and ideologically, which affect the economic and social characteristics of their participants. Analysis of these differences may itself be an important way to understand the problem of exclusion. The discussion below on exclusion and dropouts relates largely to BRAC. Thirdly, NGOs have already recognized the problems of dropouts and exclusion and have begun to explore new approaches. The BRAC IG VGD programme is one example. Fourthly, current micro credit programme strategies of achieving financial independence should result in donor funds being released and available for alternatives; one of the problems at the moment is uncertainty about what such alternatives might be. The main source of ideas may well be analysis of which poor households are benefiting from existing micro credit programmes in the context of village-wide profiles of poverty to provide some understanding of the characteristics of households that are not benefiting. There are impact assessment studies (e.g. Mustafa et al, 1996) and major research projects which have relevant findings (e.g. Khandker and Chowdhury 1995 was one of nine linked papers from a research project led by BIDS that examined three major credit programmes in Bangladesh). Understandably though, the main focus of research on micro credit programmes, until now, has been on the members themselves, and to some extent on dropouts, rather than on poor non-members who for one reason or another are excluded. However, there is some information which can be used to explore the reasons for exclusion. Evans et al (1995) have used PRA and surveys to examine barriers to entry to BRAC's Rural Development Programme (RDP).

Programme Exclusion

In discussing the issue of exclusion a basic point of departure is to distinguish between programme-driven exclusion and self-exclusion. Programme-driven exclusion is of three types. First, it may be a response by other new members at the time of group formation to perceived risk of default which, if it occurs, may affect their own access to credit under joint-liability provisions. Second, it may result from programme staff who perceive that their targets and performance indicators will be more easily attained by focusing on "good" credit risks; they may also find it easier to start operations by utilizing the services of village leaders and this may result in effective exclusion for some because of local social and political factionalism. Thirdly, it may result from the limited size of the programme; too may people interested and not enough places available. Evans et al found no evidence to support these propositions. Despite this, one frequently hears the view that, with the renewed emphasis on accountability as programmes scale up and donors have more at stake, some form of programme-driven exclusion does occur.

Self-exclusion may occur for a wide variety of reasons. Evans et al found that in their sample of over 24,000 households, 75% were eligible to be members of BRAC's RDP but less than one-third had in fact joined. Their analysis (Table 7 p.16) showed that these non-members were more likely than members to have: no adult with formal education; smaller household size—probably representing extremes of the life cycle; fewer assets; more wage labour dependence; and, lower monthly per capita income. Interestingly, neither health nor household crises were significant explanatory variables of non-membership and female-headed households were members in proportion to their presence in the sample. When target group non-members were asked about the reasons for not joining, lack of resources was the most common reason for the poorest among them and "no perceived benefit" for the better off. Social pressure was the third significant factor. They conclude that there are a group of resource constrained households, about 15% of eligible households, for whom simple expansion of the RDP is not likely to result in their participation.

Programme Dropouts Hassan and Shahid (1995) have provided an analysis of dropouts for the RDP in Matlab and Khan and Chowdhury (1995) for five RDP branches. Evans et al (1995) and Mustafa et al (1996) also provide evidence on dropouts. The first thing to observe is that much of this evidence may be of little value in understanding current patterns of dropout because the data relate to a period when substantial rationalization, improved targeting, feminization and financial tightening were underway. Expulsion (Khan and Chowdhury 1995), including some who had "graduated" out of poverty—but perhaps even before they joined BRAC-, is a better way to describe many of these cases. Here, I only consider reasons for dropout which are of current interest. The rate of dropout has steadily decreased since these studies and is now around 7%.

The Hassan and Shahid (1995) study lists 16 reasons for dropout. Four of these were related to social pressure, four to resource constraints, and four to BRAC itself. The other four were migration, death of the member, joining another NGO, and no access (as hoped) to VGD cards. The four relating to BRAC were: failure to repay loan instalments, low interest on savings, unable to count and sign, and, cancellation of membership whilst away visiting natal villages and failing to keep up weekly savings. Matlab is an area of low dropout, nevertheless there were a wide variety of reasons given for dropout. This was also found in the Mustafa et al (1996) study and they conclude that (p. 217) "The explanation for the dropout appears to be multidimensional". Specific extra programme related reasons they found included lack of prompt access to savings, too much emphasis on credit discipline, frequent policy changes and conflict among VO members. No further reasons were reported in the other two studies.

Three of the studies examined whether the economic and demographic characteristics of dropouts were systematically different to continuing members. There were few large differences though Khan and Chowdhury found less education and less land, fewer loans and a higher proportion of permanently food deficit households amongst the dropouts. Evans et al found they had smaller families and less education. Mustafa et al found that female dropouts were newer members, though with more land—on average above the maximum for the target group. Male dropouts had less wealth, smaller housing assets, less RDP savings and smaller total RDP loans. This evidence at least does not refute a hypothesis that it is poorer members who leave; it does seem to suggest that exit because of "graduation" out of poverty is less tenable. Some other anecdotal field evidence (Montgomery, 1996) does suggest that both staff and other members do seek to enforce exclusion for members who are in arrears and that these are the poorer members. One finding of Mustafa et al is that recruitment tends to be biased towards the top of the target group; this was evident for example in comparisons between a control group and new recruits. The same study showed that just under a fifth of members were not from the target group at the time of enrolment. Thus, whilst evidence comparing ex-members and current members is not conclusive, there is a suggestion that there are biases operating against poorer elements of the target group. There is evidence from elsewhere (Hulme and Mosley 1996, cited in Bennett and Cuevas 1996) that "successful institutions contributing to poverty reduction are particularly effective in improving the status of 'middle and upper income' poor". It is obviously true that the RDP procedures may not suit everybody and there was a significant relationship between those who left and membership of other NGOs in the Evans et al study. Further research is needed on the characteristics of dropouts to establish whether alternative forms of credit provision may enable them to participate.

Micro Credit and Macroeconomics

From compassion to capitalism?

The adoption in Bangladesh of the Washington agenda of market-oriented macro-economic reform promoting trade-led private sector growth seems entirely consistent with the NGO emphasis on improved rural financial services for the rural poor. The three tenets of poverty reduction in the World Bank's current discourse—labour intensive growth, human resource development and safety nets— also suggest that there is a high degree of consensus on poverty reduction strategy. The main thrust of minimalist micro credit provision is precisely intended to promote labour intensive growth. The emphasis on female users of credit and the attempt to promote women as free agents in the labour market helps reinforce the macro-economic strategy of labour intensive market-led growth. When this labour intensive growth approach is also accompanied by training, creation of better social awareness, and education and health provision, as it often is with Bangladesh NGOs, the second element of the strategy is also addressed. The third element—safety nets—is where the NGOs have traditionally been strongest but where less attention is now evident. This is not surprising. Safety nets are for those that are unable to benefit from the market opportunities created by structural adjustment (and erstwhile public sector workers and others that are "temporary" losers from structural adjustment) yet the whole thrust of the rural credit resurgence is precisely premised on the view that the rural poor are credit worthy and able to operate in the market once given the financial services denied them by traditional physical collateral based lending.

Several observers (e.g. Dichter 1996) have questioned the strength of the current donor and NGO emphasis on financial services in their poverty reduction strategies on the grounds that the approach is too narrow, that early success of some of these programmes is promoting a blueprint approach based on their model and that, as discussed above, there is evidence that not all the poor, perhaps especially the poorest, are able to benefit from this approach. In particular, there is a fear that the consequent emphasis on financial sustainability of these services is changing the ethics of the NGOs or at least creating an internal tension between "compassion and capitalism" even if it is socially-conscious. There is less space for social development and institution-building. These concerns are reinforced by the enthusiasm with which donors promote this alternative to public sector spending and mould the agendas of agencies that were, at one time, providing an alternative vision to the mainstream. The 1970s focus on structural reform and redistributivist policies is a dim memory. There are three more specific reasons, relating to the macro-economic strategy, for questioning the sustainability of this consensus.

First, the financial sector reforms are reducing the presence of traditional financial services in rural areas. The opportunities for users of micro credit to graduate to traditional services are being eroded. Secondly, the structural adjustment policies are posited on growth in (internationally) tradable goods yet many of the enterprises supported by micro credit are based on non-tradable output (e.g. animal rearing, crop processing, petty local trade itself). The relative profitability of these activities will diminish but without other complementary investments, in addition to provision of micro credit, the opportunities for poor households to diversify will remain few. Thirdly, to the extent that rural output is tradable, the opportunities for productive investment will tend to be region specific; there is likely to be greater internal differentiation in economic opportunities. This will be especially true for agriculture where marketing opportunities and agro-ecological conditions will favour some regions over others. There is strong evidence that robust non-agricultural rural growth is closely linked to local agricultural growth, principally through consumption linkages but also through backward and forward linkages in production. Many of the activities that micro credit programmes support are dependent on this rural purchasing power, from farmers especially. If this purchasing power becomes more concentrated then there will be marked regional differences in the effectiveness of micro credit services for poverty reduction.

The macro-economic reforms are only just beginning to be put in place. Whilst the stabilization measures have largely been implemented the supply side reforms have been stalled in several sectors. As they are implemented, these concerns will be of increasing relevance. It is difficult to assess how strong their effects will be—there is need for research—but they do at least suggest caution in assuming that the current consensus is here to stay.

Chapter 3

FINANCE FOR THE POOR OR POOREST? FINANCIAL INNOVATION, POVERTY AND VULNERABILITY

David Hulme and Paul Mosley

Introduction

The last decade has witnessed an explosion in both the numbers and the scale of organisations providing very small loans (micro-credit) to poor people to help them escape poverty. The Grameen Bank has spearheaded this strategy and by mid-1996 had more than two million borrowers and was advancing loans of more than US$30 million each month. Its model has been copied by many non-governmental organisations (NGOs) so that almost 25 per cent of poor rural households in Bangladesh now have access to institutional credit. Further afield the Grameen Bank is being 'replicated' in Asia (Malaysia, the Philippines, Indonesia, Nepal, China and Sri Lanka), in Africa (Kenya, Malawi, Nigeria) and North America (Canada and the USA). In Latin America the ACCION network has affiliate financial organisations in most countries operating schemes partly based on Grameen Bank principles. With the UK's Know How Fund currently exploring the possibility of using micro-credit as a self-employment strategy in Eastern Europe and the former Soviet Union, the 'movement' may soon cover the globe.

While there is growing evidence of the ability of micro-credit to reduce poverty a growing number of researchers (see Rutherford and Wright in this volume; Rogaly 1996; Bundell 1996) and practitioners (ACORD, Action Aid, Aga Khan Foundation, BURO Tangail,

Christian Aid, OXFAM, SANASA and WomanKind) are arguing that what the poor need is micro-financial services (micro-scale short and long-term savings, investment loans, consumption loans and, perhaps, insurance). A micro-finance approach can also aid institutional financial viability (Robinson 1992). This chapter examines the contribution that micro-credit and micro-finance can make to the alleviation and removal of poverty. It draws on the materials from a study of 13 financial institutions (see Appendix 1) in 7 countries (Hulme and Mosley 1996). The first part explores the concept of poverty to identify the criteria that could be used to assess poverty impacts. Subsequent sections analyse the impacts of micro-finance initiatives on income poverty, income vulnerability and groups that suffer particularly high levels of economic and social deprivation.

The conclusion argues that micro-credit has proved effective in poverty reduction but has done little to help the poorest. A shift towards a micro-financial services approach is needed to permit financial innovations to more effectively meet the varying needs of the poor and poorest. While micro-finance should be an element of poverty-reduction strategies it is no panacea. The contemporary micro-finance bandwagon (the inaptly named Consultative Group to Assist the Poorest and the Micro-credit Forum of 1997 amongst other initiatives) should not obscure the fact that poverty-reduction requires many other forms of action.

Poverty, Vulnerability and Deprivation

The definition of what is meant by 'poverty', how it might be measured and who constitute 'the poor' are fiercely contested issues. At the heart of the debate about defining poverty stands the question of whether poverty is largely about material needs or whether it is about a much broader set of needs that permit well-being (or at least a reduction in ill-being).[1] The former position concentrates on the measurement of consumption, usually by using income as a surrogate. Although this approach has been heavily criticised for its 'reductionism' and 'bias to the measurable' (Chamber 1995) it has considerable strengths in terms of creating the potential quantitatively to compare and analyse changes in the access of different

---

[1] For detailed discussions of the definition of poverty, see Chambers (1983 and 1995), Doyal and Gough (1991), Greeley (1994), Ravallion (1992) and Townsend (1993).

people to their most pressing material needs (Townsend 1993). Greeley (1994: 57) has strongly defended the use of income-poverty measures: 'an absolute and objective poverty line is a form of information that empowers the poverty reduction agenda and encourages appropriate resources allocations'.

The case is far from absolute, however, and even adherents acknowledge that 'there is a broad agreement that income is an inadequate measurement of welfare' (*Ibid*: 51). Chambers (1983, 1995) has recorded the many forms of deprivation that very poor people identify themselves as experiencing that are not captured by income-poverty measures. These include vulnerability to a sudden dramatic decrease in consumption levels, ill-health and physical weakness, social inferiority, powerlessness, humiliation and isolation. Such dimensions of poverty are significant in their own right and are also essential analytical components for the understanding of income poverty. The failure of income measures to capture such deprivations can be illustrated at the macro level by the 'weakness in the correlations between income-poverty and some other deprivations' (Chambers 1995: 12).[2] At the household level it is illustrated by Jodha's (1988) finding that households in Rajasthan who became income-poorer over the period 1963-6 to 1982-4 regarded themselves as being better off in terms of their self-defined criteria of the quality of their lives.

Although the debate about reductionist or holistic approaches is commonly presented as an 'either... or' argument this need not be the case. In the materials that follow an integrated approach is adopted. While we believe that the adoption of such a broad approach is essential, if the impact of financial innovations on poor people is to be better understood, the reader should recognise the analytical complexity that this introduces. Most obviously, having multiple criteria makes it less easy to judge an innovation or service simply as 'good' or 'bad'. Instead, the likelihood arises of institutions reducing poverty and deprivation in some respects while being neutral or negative in others. Thus policy implications may need to be qualified, rather than dropping as neat messages out of the analysis.

---

[2] 'Strikingly, the latest *Human Development Report* (UNDP 1994: 15) shows Sri Lanka, Nicaragua, Pakistan and guinea all with per capita incomes in the $400-500 range, but life expectancies of, respectively 71, 65, 58 and 44, and infant mortality rates of, respectively, 24, 53, 99 and 135' (Chambers 1995: 12).

The definition of poverty and deprivation is not merely of analytical significance: it also has a strategic dimension. A concentration on poverty as 'income poverty' is usually associated with a conceptualisation of poverty-reduction as moving households from a stable 'below poverty line' situation to a stable 'above poverty line' situation. This leads to a focus on promotional strategies 'raising persistently low incomes' (Dreze and Sen 1989: 60-1) which, in terms of financial services, emphasise (often exclusively) the provision of credit for income-generation through self-employment. By contrast, a broader view of poverty that conceptualises income levels as fluctuating below and perhaps above a poverty line suggests the dampening of dramatic reductions in income (and other entitlements) as a means for poverty reduction and introduces quite different strategic emphases. Protectional strategies (*Ibid*) become significant: in terms of financial services this fosters a focus on voluntary savings mechanisms, emergency consumption loans and relatively low-risk income-generation activities that are unlikely to create indebtedness.[3] Distinguishing between promotional and protectional approaches does not require that they are seen as unrelated or as competing directly against each other. Effective promotional strategies, that raise household incomes and create additional assets, can make the protection of a minimum standard of living much easier. Conversely, effective protectional strategies may permit households to undertake investments that they had previously regarded as being too risky.

Two further points must be noted. The first is that, contrary to much recent writing, micro-enterprise and/or small-enterprise development should not be equated with poverty-reduction. At times the two will coincide but this needs empirical validation and should not be assumed, as is commonly the case. The second point, related to the first, is that most of the poor households that we have studied cannot be viewed simply as micro-enterprises, self-employed poor or labouring poor (for example, see Remenyi 1991: 8). The vast majority of households that we have studied are 'foxes' not 'hedgehogs' (Chambers 1995: 23). They do not have a single source of livelihood support: rather, depending on season, prices, health and other contingencies, they pursue a mix of activities that may include

---

[3] Although concerns about 'protection' are associated with writers of the left, the concern has also been raised by those on the right. For example, von Pischke's (1992) contention that 'credit' should be called 'debt', and that being indebted may be very risky for poor people, is based on similar premises.

growing their own food, labouring for others, running small production or trading businesses, hunting and gathering, and accessing loans and subsidies (from the state, friends or NGOs). In terms of economic behaviour, they are closer to the manager of a complex portfolio than the manager of a single-product firm.

Finance for the Poor: Improving Incomes

The primary process by which financial services are envisaged as reducing poverty is by the provision of income-generating loans. According to Muhammud Yunus of the Grameen Bank a virtuous circle can be established: 'low income, credit, investment, more income, more credit, more investment, more income' (quoted in International Development Support Services 1994: 6). As is discussed below, this notion of sustained growth in income, production, credit and investment captures a part of the experience of poor households that borrow, but only a part. The differing abilities of borrowers, their initial economic endowments and social positions and the wider economic environment (and its fluctuations) ensure that no simple model can explain the complex empirical findings of our research.

The obvious starting point in examining the impacts of credit on the incomes of the poor is to see the degrees to which they can access loans (Table 1). All our case study institutions, with the exception of the Kenya Industrial Estates—Informal Sector Programme (KIE-ISP) make loans to some people with incomes below national poverty lines. As one would anticipate, institutions that target the poor have much higher rates of participation by the poor than open-access schemes. The Thrift and Credit Cooperatives in Sri Lanka (SANASA) are the only non-targeted scheme that manages to incorporate a majority of poor members. Only the Bangladesh Rural Advancement Committee - Rural Development Programme (BRAC-RDP), Grameen Bank, Bank Rakyat Indonesia (BRI) unit desas, Badan Kredit Decamatan in Indonesia (BKK) and SANASA have reached coverage rates that make it possible for them to have significant impacts on poverty at the meso and micro-level. Notably, three of these institutions are non-targeted, illustrating that mass coverage open-access schemes, with appropriate policies,[4] may play

---

[4] Important amongst these are the 'economic' policy of tapered interest rates (BRI's unit desas and BKK) and the 'social' policy of encouraging low-income borrowers (SANASA).

a role in poverty-alleviation despite the preference of many donors for targeted interventions. The income impacts recorded (Table 1) are only a snapshot of a constantly changing situation and evidently different schemes are achieving quite different results. Nevertheless, three general points can be drawn.

1. Well-designed lending programmes can improve the incomes of poor people and for a proportion of cases can move the incomes of poor households above official poverty lines in large numbers (BRAC's RDP, Grameen Bank, BRI's unit desas, BKK and SANASA).

2. There is clear evidence that the impact of a loan on a borrower's income is related to the level of income (Figures 1 and 2). For rural Asia the Bangladeshi cases (BRAC and the Thana Resource Development and Employment Programme—TRDEP) illustrated this most clearly and the available data on Sri Lanka and Indonesia confirmed this picture. This finding should not be unexpected given that those with higher incomes have a greater range of investment opportunities, more information about market conditions and can take on more risk than the poorest households without threatening their minimum needs for survival. Such data confirm the argument made in our Bangladeshi case study (Montgomery, Bhattacharya and Hulme 1996) that income-generating credits are not 'scale neutral', but have differential utilities and effects for different groups of poor people. This is an important finding as it indicates that:

   a. credit schemes are most likely to benefit the incomes of what may be termed the 'middle' and 'upper' poor

   b. the poorest or 'core poor' (see Rahman and Hossain 1992) receive few direct benefits from income-generating credit initiatives and so alternative assistance strategies (in the finance and other sectors) need to be developed

   c. institutions seeking to provide income-generating credit to the poor, while pursuing their own financial viability will have a tendency to concentrate on the 'upper' and 'middle' poor[5] (see Hulme and Mosley 1996, Chapter 8).

---

[5] Our study of Bangladesh revealed this to be the case for TRDEP and to be the direction that BRAC's RDP has taken in the 1990s. Evidence of the Grameen Bank's problems in working with agricultural labouring households and other studies (IDSS 1994) indicate that it is also operating in this niche.

Table 1: Finance for the Poor: Impacts on Income for Case Study Institutions

Organisation	Target group	Proportion of borrowers below poverty line	Number of borrowers (1992)	Status of sample	Average borrower income as % of poverty line before last loan	IMPACT of last loan on borrowers income (real terms)	% of borrowers crossing poverty line	Estimate of numbers crossing poverty line in 1992	Other information
BRAC-RDP, Bangladesh	Poor especially	Vast majority	650,00	Completed 1st loan Completed 3rd loan	68 (117)[1] 58 (99)[1]	1.1% increase 6.4% increase	– –	– –	Evidence that it is moving away from a focus on the core poor
TRDEP, Bangladesh	Poor families	Vast majority	25,000	Completed 1st loan Completed 3rd loan	80 (136)[1] 103 (197)[1]	23.0% increase	–	–	Lends to non-poor and poor households, but not core poor
Grameen Bank, Bangladesh	Poor, almost exclusively women	Vast majority	1,400,000	(see Hossain 1988)	–	Members have incomes 43% above target group in control villages, and 28% above non-members in target groups of assisted villages	–	–	Difficulties in working with core poor-agricultural labourers are 60% of target group but only 20% of membership

(Contd.)

104 *Who Needs Credit?*

(Continued)

Organi-sation	Target group	Proportion of borrowers below poverty line	Number of borrowers (1992)	Status of sample	Average borrower income as % of poverty line before last loan	IMPACT of last loan on borrowers income (real terms)	% of borrowers crossing poverty line	Estimate of numbers crossing poverty line in 1992	Other information
BancoSol Bolivia	No	29%	–	Half about to borrow, half 6 or more loans	480	91% of borrowers borrowers had increase in income; 89% of borrowers below poverty line experienced income increase greater than 50%	8.0%	–	To achieve financial viability BancoSol is promoting larger loan sizes; this is likely to further distance it from the poor
BRI unit desas, Indonesia	No	7%	2,400,000	See volume 2, chapter?	195	20.7% (1990)	8.4%	48670	Only 7% of borrowers poor but BUD has wide coverage
BKK, Indonesia	No	38%	499,000	See volume 2, chapter?	163	Poor borrowers increased income by more than 18%	6.6%	36932	–
KURK Indonesia	No	29%	158,000	See volume 2, chapter?	168	Poor borrowers increased income by more than 2%	3.3%	3753	–

(Contd.)

Finance for the Poor or Poorest? 105

(Continued)

Organi-sation	Target group	Proportion of borrowers below poverty line	Number of borrowers (1992)	Status of sample	Average borrower income as % of poverty line before last loan	IMPACT of last loan on borrowers income (real terms)	% of borrowers crossing poverty line	Estimate of numbers crossing poverty line in 1992	Other information
K-REP Juhudi,	Small entre-preneurs	–	2,403 (1992)	Completed 1st loan Completed 1st and 2nd loan	–	Associated with 24% increase in income Associated with 44% increase in income (both loans)	–	–	Evidence of increased incomes and assets (see Hulme & Mosley (1996) volume 2)
KIE-ISP,	Not the	0%	1,700 (1992)	Completed 1st loan Completed 1st and 2nd loan	–	Substantial increases in enterprise profitability	0%	0%	No direct poverty impact-' second round' effects only
MMF, Malawi	Poor especially women	Vast majority	223	–	–	Evidence on average MMF borrowers increase income	–	–	Scheme in experimental phase only
SACA, Malawi	Small farmers	48%-7%[2]	400,052 (1992)	–	–	High variable	2%	–	SACA has has had involvement with the core poor

(Contd.)

(Continued)

Organisation	Target group	Proportion of borrowers below poverty line	Number of borrowers (1992)	Status of sample	Average borrower income as % of poverty line before last loan	IMPACT of last loan on borrowers income (real terms)	% of borrowers crossing poverty line	Estimate of numbers crossing poverty line in 1992	Other information
SANASA, Sri Lanka	Rural people–no specific target group	52%	702,000 District	Kurunagala District Monaragala	192 154	26.0% increase 18.9% increase poverty	} 25% of those } below } line	} 30,000 }	Increased incomes especially significant for middle-income households

Notes: 1. As a percentage of the 'core poor' poverty line (Rahman and Hossain 1992).
2. Core poor.

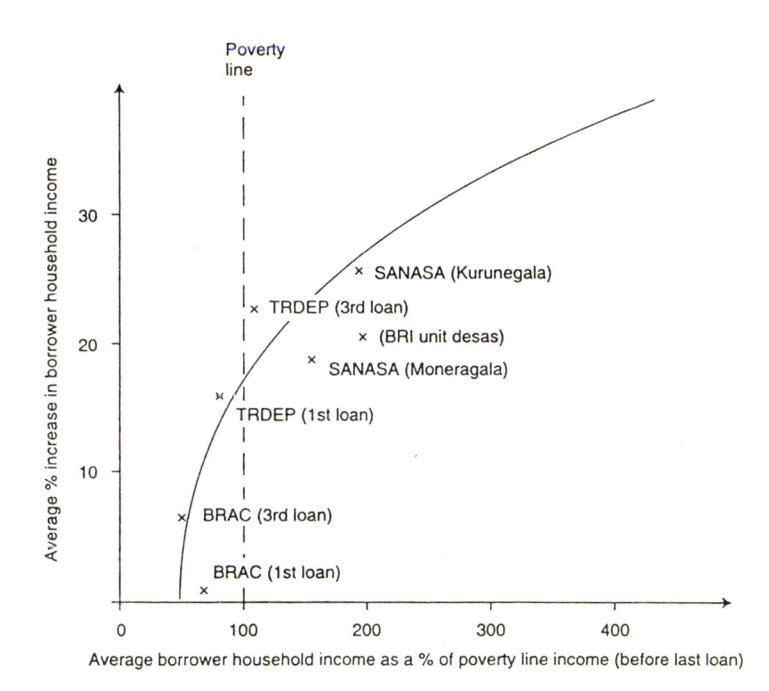

Fig. 1: The Relationship of Average Borrower Income to Average Increase in Household Income since Last Loan

The impacts of income-generating credits in the medium term cannot be simply understood in terms of a promotional model of credit, investment and income. A significant minority of investments fail (leading to decreases in income) while many investments that increase income soon reach a plateau (for example, operating a rickshaw, manually hulling rice, adopting HYVs and inputs on a small farm). For the latter, credit schemes give borrowers an important 'one step up' in income, however, 'survival skills' rarely provide the technological or entrepreneurial basis for poor borrowers to move on to the 'escalator' of sustained growth of income. During fieldwork in Bangladesh, Kenya, Malawi and Sri Lanka respondents persistently commented on their desire for lending agencies to develop new forms of income-generation that would permit borrowers to move beyond their present levels of income. Even for urban-based BancoSol (for which the most dramatic increases in income were

108 *Who Needs Credit?*

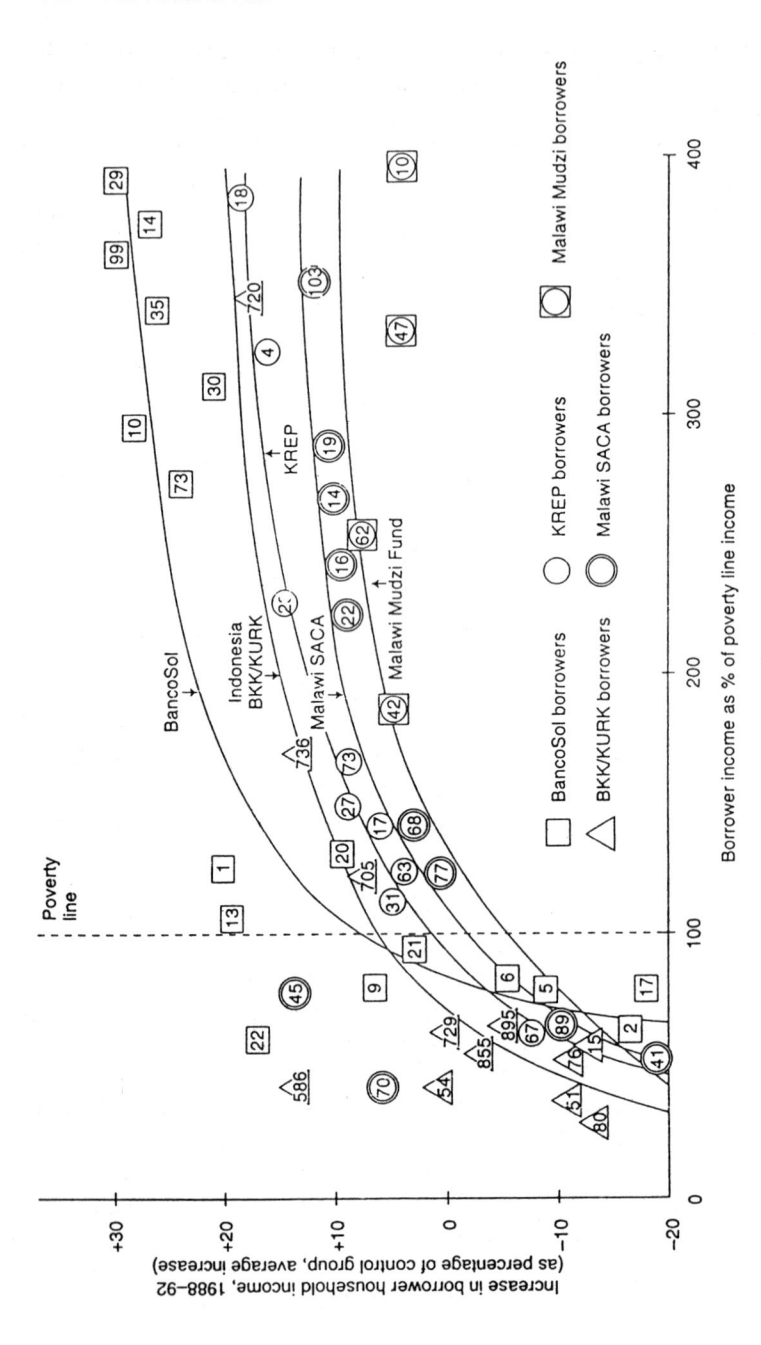

Fig. 2: Loan Impact in Relation to Borrower Income: within-Scheme Data

recorded) only 11 per cent of long-term borrowers had been able continuously to achieve income growth, while at least 41 per cent of borrowers had (at best) reached a static income situation.

In sum, while our study confirms the emerging consensus that well-designed micro-credit schemes can raise the incomes of significant numbers of poor people it also indicates that such schemes are not the panacea for poverty-reduction that some claim. There are trade-offs between the goals of poverty-reduction and institutional performance, and credit has differential impacts on different groups within 'the poor'.

Finance for the Poor: Reducing Vulnerability

Poor people suffer not only from persistently low incomes but also from 'the precarious nature of their existence, since a certain proportion of them undergo severe—and often sudden—dispossession, and the threat of such a thing happening is ever present in the lives of many more' (Dreze and Sen 1991: 10). Poverty, in this sense, involves both the fear of and the operation of events that drive down existing levels of income and consumption. Such events include the illness or death of a member of the household, medical expenses, funeral costs, crop failure, the theft of a key asset, a dramatic change in prices or the payment of dowry. Often such events are linked together, as illustrated by a Malawi Mudzi Fund (MMF) member in Malawi: 'I failed to repay because my business failed because my spouse fell ill and I had to pay for a doctor, and there was little money because we had to buy more food than usual because our crop was spoiled by the drought'. Can financial services reduce the vulnerability of such households so that such ratchets are less likely to operate?

The evidence available from our case studies reveals their relatively limited contribution to reducing the vulnerability of poor households to a sudden dramatic decline in income and consumption levels (Hulme and Mosley 1996: 115-7). The schemes explicitly targeted at 'small enterprise' and 'small farms' (KIE-ISP and Malawi's Smallholder Agricultural Credit Administration (SACA)) basically bypass vulnerable households because of their loan screening and membership procedures. If they impact on vulnerability then this must be through second-round effects on the labour market. Such effects were very limited for our case studies (*Ibid*: Chapter 4). The main contribution of the non-targeted programmes (BKK, BRI's unit desas, KURK and SANASA), all of which offer services to the rural population in general terms, is on the savings side of the financial market. These schemes provide relatively low transaction cost savings services that permit vulnerable households to 'store cash', which earns interest but which can be rapidly accessed in times of crisis. SANASA's Federation-level policies to shift its services more towards the needs of poorer people have led to it extending its services into loans for very poor households, particularly for small seasonal agricultural production credits and for 'instant' short-term consumption loans (Hulme, Montgomery and Bhattacharya 1996). Both of these activities extended its contribution to the vulnerable poor beyond those of the Indonesian institutions.

The most effective interventions might be expected to come from programmes that exclusively target the poor, but here again the evidence is mixed. BRAC, the Grameen Bank and MMF all managed to provide credit to core poor households, whereas TRDEP, while having 79 per cent of its borrowers in the 'poor' category, systematically avoided the very poor. Its membership was focused around the 'poverty line' of ownership of 50 decimals of land, so that it did not work with day labourers, the landless or widows and divorcees who headed households. BRAC, Grameen Bank and MMF all provided income-generating credits to a proportion of very poor people who on average raised their incomes and assets during the course of these loans. Worryingly, both BRAC-RDP and the Grameen Bank recently appear to be moving away from working with significant proportions of the core poor and focusing their activities on the middle and upper poor, rather than the most desperate. The institutions that target 'the poor' have generally performed less effectively on the savings side of the market: whilst both BRAC and Grameen Bank have generated large volumes of savings these funds are not easy to access when member households find themselves facing a crisis, and members are very critical of this situation.

Rahman and Hossain (1992) propose that the vulnerability of the poor can be understood in terms of a set of 'downward mobility pressures'. These are

- structural factors within the economy, particularly demand for labour, demand for the products and services of poor people and seasonality

- crisis factors, such as household contingencies or natural disaster, and
- life-cycle factors, particularly the proportions of economically active and dependent persons in a household.

What contribution have the case study institutions made to reducing downward mobility pressures and strengthening household security in terms of this framework?

Downward Mobility Pressures 1: Structural Factors

In terms of structural factors then our case study institutions have had, at best, only a marginal impact. The data available on the impact of loans on employment outside the family are very limited and, where it occurs, it is associated with a rapidly growing formal economy (such as Indonesia) and with lending to the non-poor or less-poor. While the main employment gains were intra-family, these were also small, ranging from 1.9 per cent for 'lower-impact' schemes to 5.6 per cent for 'high-impact' schemes (Hulme and Mosley 1996: 90). There are thus no grounds for believing that in any of our case study countries innovative financial services have sufficient impact on demand for labour at regional or local levels to lead to raised wage rates.

The impact of credit-financed activity on the demand for the products and services of borrowers has been examined by other writers, particularly with regard to Bangladesh. Osmani (1989) has pointed out that in a rural economy in which agricultural output is growing at a very slow rate the demand for the goods and services produced by poor people's enterprises—which are not traded internationally and are only marginally traded into urban centres—is likely to be a constraint on both the volume of production and returns. For some sectors loan-induced activity may expand supply to a position at which 'the rate of return [is] below the cost of borrowing' (*Ibid*: 16). This factor might well explain why the Grameen Bank and BRAC have significant numbers of dropouts each year. Even those who have queried Osmani's gloomy prognosis accept that in many lending sectors demand is a fundamental constraint on loan expansion and enterprise returns and that much of the income generated by trading activities might be simply a redistribution of income from existing traders to new traders (Quasem 1991: 131).

So, as with employment, there is no evidence that structurally based constraints on demand for the products and services of the poor are likely to be removed by credit-induced activity: rather, they are dependent on changes in the wider economy.[6]

With regard to seasonality there is only limited evidence that any of our rural-focused case study institutions have been able to stimulate new activity in the 'off-season'. SACA's lending programme focuses only on peak-season activities. TRDEP, the Indonesian institutions, MMF and SANASA rely upon borrowers using their 'survival skills' to identify investment opportunities and do not supply technical assistance. For all these cases the basic mode of project identification by borrowers could be described as 'copying' what others are doing. One of the main causes of the initial failure of MMF related to the seasonal problems (drop in demand, need to divert income-generating loans to food purchases) encountered by borrowers in the period preceding the harvesting of the maize crop. The Grameen Bank and BRAC have financed research and development operations in an attempt to diversify investment opportunities, and some of these R&D initiatives have focused on activities that generate employment and income in the 'off-season'. Although the Grameen Bank has encountered difficulties in its R&D work (particularly for deep tube-wells, fish-farming and the use of HYV rice), its loans have stimulated large increases in milch cow ownership amongst members. In 1991, 12 per cent of Grameen Bank loans (140,317) were for milch cows, creating the possibility for an income flow during the troughs of the agricultural cycle. BRAC-RDP has promoted sericulture and poultry production through large-scale training and technical assistance programmes and is increasingly lending for these activities. Such initiatives have created opportunities for new income flows during the lean months of mid-September to mid-November. It must be noted, however, that the costs of providing borrowers with training does add significantly to overall operational costs. So, for those institutions involved in research and development, there is some evidence of a capacity to enhance the income security of a minority of borrowers in the lean season. However, for most institutions, and the vast majority of borrowers,

---

[6] For example, in Moneragala, many SANASA members were benefiting from loan-financed diversification into sugar cane production. The creation of this opportunity lay in national agricultural policy, however, rather than financial service provision.

seasonal problems of low demand for labour, products and services have not yet been reduced by the provision of financial services.

Downward Mobility Pressures 2: Crisis and Contingency

Turning now to downward mobility pressures caused by crisis (sickness leading to loss of earnings and medical bills, theft of assets, flooding, drought) a number of mechanisms by which financial services, and loan-financed activity is associated with an increase in the assets controlled by vulnerable households. This extends crisis-coping mechanisms as the possibility of selling or pawning assets, to bridge a consumption deficit, is increased. For BRAC, Grameen Bank and SANASA for which increases in assets were reported and significant numbers of poorer households were borrowers— there is some evidence of asset growth providing an additional buffer against contingencies.

A second way in which financial services, and in particular loans, might reduce crisis-related vulnerability would be by encouraging borrowers to invest in risk-reducing technologies or income-generation activities that have a lower risk than pre-existing activities. The identification of risk-reducing technologies is difficult because of differing contexts, but the most obvious is irrigation which permits cropping when rainfall fails. In the drought-prone areas covered by our study (Malawi and the Moneragala District of Sri Lanka) the loan size and conditions of the institutions studied (SANASA, SACA and MMF) were insufficient for a 'lumpy' investment such as tube-well and pump. In Bangladesh[7], both the Grameen Bank and BRAC have experimented with group loans for deep tube-wells (DTWs). Both agencies have encountered severe difficulties and have withdrawn from such activity: the Grameen Bank now directly manages its DTWs, while an evaluation of BRAC's 70 DTWs has revealed that only a small proportion are viable and that many borrowers, and particularly the poorer, have lost assets or have suffered income reductions because of involvement in such activities (van Koppen and Mahmud 1995). In effect, the poor have absorbed the risks of the DTW experiment.[8]

---

[7] It is probably more appropriate to regard irrigation in Bangladesh as productivity-enhancing rather than productivity-securing.

[8] For a discussion of an apparently more successful approach with shallow wells and other forms of irrigation, see Wood and Palmer-Jones (1991).

The question of whether loans are associated with encouraging borrowers to operate in lower-risk activities is complex. According to the conventional wisdom of finance, borrowers seeking to increase their income will take on additional risk, as returns on investment are partially a function of increased exposure to risk. If this is the case, then lending to poor people is likely to raise their average incomes but will, for a proportion of borrowers, increase the likelihood of a crisis because of 'business failure'. Although we do not have systematic data on such incidences, considerable evidence was gathered during fieldwork that a minority of borrowers become worse off because of borrowing; that is credit from a case study institution increased their vulnerability. This situation is vividly illustrated by the case of a TRDEP borrower documented in a report to donors (Box 1). There is also much additional evidence:

- BancoSol staff estimate that 10-15 per cent of borrowers' enterprises go bankrupt (Mosely 1996);
- BRAC borrowers reported the forced seizure of defaulters' assets (such as livestock and cooking pots) and their sale in order to cover the costs of loans that members had not repaid (Montgomery, Bhattacharya and Hulme 1996). A more dramatic example is provided by Khan and Stewart (1992) who write of a conversation with BRAC women in which they described pulling down a member's house because she had not paid back a loan;
- 20 per cent of TRDEP members drop out before their third loan —'usually the most vulnerable who drop out as a result of failing to use the loan successfully' (IDSS 1994: 3);
- For the Kenya Rural Enterprise Programme (K-REP) Juhudi scheme, peer pressure is reinforced by the 'unofficial' pledging of assets within groups (though this is known by K-REP staff). Some borrowers have forfeited significant assets to other members when loan-financed activities failed;
- Around 25 per cent of MMF-financed activities (in its early phase) failed—scheme planners and managers misunderstood the nature of borrower risks;
- At the time of survey a growing number of KIE-ISP borrowers were being classified as Class D (i.e. non-viable) on the organisation's A to D scale for investment risk. Many are likely to be taken to court or lose key assets in the future.

> **BOX 1**
>
> **When a Loan goes Wrong: Driving into Debt**
>
> Kawser Ahmad was a hard-working young day-labourer, married, with two small children. Because he was the cousin of a prosperous bazaar trader who became a TRDEP *kendra* Chief, he found himself a member of his cousin's *kendra*. His wife did not join him, because the group animator (GA) felt that, as a poor man, Kawser would not be able to make more than the smallest weekly repayment. His cousin advised him to buy a rickshaw-van, which he did, but because his loan was 2,500 taka, he had to borrow another 500 taka from his cousin in order to buy a rather run-down machine. Kawser had never ridden a van before, and was also very cautious about giving up his day-labouring work. He therefore tried to combine the two, which meant hiring the van out on days that he got work in the fields. This didn't work—he had trouble hiring the van out to drivers who had not certainty of access to the van every day. So he gave up that strategy and tried full-time driving himself. This worked more or less well for a few months, but then the rains came and customers fell away. This coincided with a lean time for day-labouring, so for some weeks his family got by through eating less and borrowing. There was no question of keeping up with the loan repayments, so he had to borrow again from his cousin. At the end of the loan cycle the other members removed him from the *kendra*. He was obliged to sell the van—at a considerable loss—to settle his debts.
>
> Source: IDSS (1994: 65)

While our study focused on new and continuing borrowers, the findings point to the need for further research examining why borrowers leave innovative financial institutions and if such departures are associated with increased levels of debt or reduced asset levels. Given the scale of 'dropping out' (15 per cent per annum for the Grameen Bank, which is 300,000 members a year; 10-15 per cent per annum for BRAC, or 181,700 members, in 1992 and 1993) there well may have been significant under-reporting of credit-induced crisis in most studies of finance for the poor. The virtuous circle of 'low income, credit, investment, more income, more credit, more investment, more income' is seductive; unfortunately, it does not mirror the reality that the majority of very poor households face in sustaining a livelihood.

A third way in which financial services may reduce vulnerability is by providing facilities for very poor households to 'store' windfall and seasonal cash and earn a return. The institutions we studied had very different approaches to savings. At one extreme stand SACA and the KIE-ISP, which only provide loans. At the other stand the BRI unit *desas* and SANASA in which savings activities are as important as lending activities. The unit *desas* have proved that the main demand for financial services from low income Indonesians is for safe saving facilities. Similarly, in Sri Lanka, SANASA's experience illustrates that the highest priority of poorer households for financial services is for easy-withdrawal saving programmes. In between these two positions come BRAC[9], the Grameen Bank, TRDEP, K-REP and BancoSol: all of these have compulsory savings programmes but limit member access to these savings.

The final mechanism by which financial institutions can reduce the effects of crises on vulnerable households is by extending the entitlements[10] of such households when a crisis occurs. The only institutions which systematically offered such services were SANASA in Sri Lanka through its 'instant loans' service and the Grameen Bank through its Group Fund and Emergency Funds[11]. The SANASA facility provides members with access to a Rs. 500 loan for a three-month period, usually within 24 hours of application, for any purpose. Although conditions vary from village to village, the usual requirements are than an applicant has a savings account, has a good repayment record from earlier loans and pays interest at a rate of 5 per cent per month (more than three times the usual rate). Such loans are in high demand from the poorer members of SANASA groups who reported that instant loans permit them to cope with emergencies (such as medical bills, lack of food and meeting the costs of essential social obligations) at lower cost than alternative coping mechanisms (such as mortgaging crops, land or labour, a distress sale of assets or borrowing from a trader). Our focus on income-generating loans meant that we did not gather sufficient data to comment quantitatively on the effects of consumption loans.

---

[9] BRAC has experimented with a pilot voluntary savings scheme that permitted savers rapid and easy access. An evaluation found that members valued this service highly and that the scheme could contribute to branch financial viability. For unknown reasons BRAC senior management has decided not to extend this initiative.

[10] Entitlement refers to the 'set of alternative bundles of commodities' over which a person (or group of persons) can establish command through ownership or rights of use. For a detailed discussion, see Sen (1981).

[11] Several of the other institutions studied unwittingly provided loans for consumption when borrowers used income-generation credit for non-approved purposes.

Nevertheless, there was a substantial 'voice from the village' that such loans are a significant addition to the survival strategies of vulnerable households.

The Grameen Bank's 'group fund accounts' serve a similar role for Grameen Bank members, although in this case members borrow from a portion of their own forced savings. The compulsory 1 taka per week savings and 5 per cent group tax that members pay on each loan are credited to a group fund account. Group members can borrow up to half of the funds in this account—with group approval. In 1991 some 162,000 such loans were made to a total amount of Tk. 189 million: 59 per cent of these loans were for social, household or medical needs (Grameen Bank 1992: 102). While these loans clearly benefit many Bangladeshi households, Bank members report that group funds are not as accessible as they would like. The Bank also operates an 'Emergency Fund' financed by a 5 per cent charge of all loans over Tk. 1,000. This is intended to cover death, disability, accidents and crisis-induced defaults. Although it has accumulated to Tk. 145 million, disbursements have been minimal (Fuglesang and Chandler 1993: 106) and, consequently, the fund is unpopular with members, being seen as a disguised borrowing charge[12].

The SANASA and Grameen Bank schemes both illustrate the possibility of financial services extending the entitlements of poorer households when contingencies occur. Although both schemes require further development (and detailed study of who accesses them and for what uses), other institutions could well enhance their poverty-alleviating impacts by the introduction of small-scale, rapid-access consumption loans.

Downward Mobility Pressures 3: Life-cycle Factors

This study sheds little light on the ways in which innovative financial services may help or hinder households to cope with the stresses of demographic change in household structure. This is for two reasons. The first is lack of data, as any detailed commentary on such matters would demand longitudinal information beyond the resources of our research. The second relates to the sheer complexity of such factors. This can be illustrated by the findings of a recent

---

[12] Interestingly, Grameen Bank did use emergency loans for food purchase in the Rangpur area following severe floods in 1991; 18,000 members were loaned Tk. 300 each, from Grameen Bank funds, to alleviate their hunger.

study of the impacts of Grameen Bank and BRAC on contraceptive usage (Schuler and Hashemi 1994). It found that participation in the Grameen Bank's credit programmes was associated with dramatic and statistically significant changes in contraceptive behaviour (59 per cent usage for bank members compare to 43 per cent usage for a matched control group in 1992). Such a situation is almost certainly likely to lead to Grameen Bank members having smaller families than non-members. In theory, this should lead to Grameen Bank members having lower household dependency ratios and thus having more resources for investment, income enhancement and asset growth. In the longer term, though—given the fact that the elderly in rural areas rely heavily on the support of their sons—these benefits may be outweighed by having fewer sons (or no sons) to provide support during old age. The possibility of greater 'child quality' in smaller families (better education, better nutrition, better health, and so on) might partially offset this if the children of smaller families ultimately derive higher incomes (and use a proportion of these higher incomes to help their parents in old age).

What is clear, however, is that none of our case study institutions generates a sufficient income for more than a handful of its poor borrowers to amass an asset base that could generate a quasi-pension when they are too old or infirm to operate their own income-generating activities. For the 'micro-entrepreneurs' of developing countries security in old age remains a function of social relationships with family, kith and kin and 'neighbours' rather than an earnings-financed private entitlement.

Finance for the Poor: Influences on other Forms of Deprivation

Poverty is not purely about material conditions. It also refers to other forms of deprivation, and in this section the effects of innovative financial services on those who suffer from social inferiority, powerlessness and isolation[13] are considered. These variables are

---

[13] Chambers (1995) also includes 'humiliation' in his conceptualisation of deprivation. While we concur with this inclusion, the nature of our field research did not allow us to capture much data on such a sensitive and methodologically difficult item. However, it was clear that in some circumstances credit groups do humiliate members who cannot meet their loan obligations and that field staff are insensitive. For example, one BRAC female borrower described how a programme assistant confronted her father-in-law when she was unable to meet instalments for a failed project and subsequently how she was stigmatised within her household.

closely interlinked (with each other as well as with income and vulnerability) and so we examine them together in relation to three groups who commonly experience relatively intense levels of deprivation and discrimination: women (particularly female-headed households); the very poor (particularly the assetless, the landless and those dependent on agricultural labouring); and the disabled.

Women, Women's Empowerment and Gender Relations

In all case study countries significant female-male gaps occur, indicating the unequal economic and social relations between women and men (UNDP 1994). Giving women access to credit, it has been argued, is a means by which both their economic standing within the household and social position within society can be improved. This argument has been particularly significant in Bangladesh where women's position is so poor *vis-à-vis* men, and where female participation in credit schemes has now reached very high levels (Hulme and Mosley 1996: 126-7). Two forces have been particularly significant in increasing women's involvement in credit programmes. The first relates to the financial viability of institutions. As revealed in our case studies and the work of others[14], in many contexts female borrowers have proved more reliable than male borrowers: consequently, some lenders have found that their financial performance can be improved by focusing on female borrowers. The second force stems from aid donors who 'discovered' women in the early 1980s and have subsequently encouraged recipient agencies to provide women with more assistance. For credit this has meant pressures to increase the proportion of female borrowers, to at least 50 per cent, and sometimes to focus exclusively on women. Behind the belief that 'loans for women' will lead to their economic and social advancement are a number of assumptions: that women will use loans for their own enterprises; that these enterprises will be successful; that women will control the profits derived from such enterprises; and that greater involvement in economic activities will strengthen the social and political position of women in society.

---

[14] For the Grameen Bank in 1985 81 per cent of female borrowers had no overdue repayment instalments as against 74 per cent for men (Hossain 1988: 32). For the Malawi Mudzi Fund in late 1990, the on-time repayment rate for women was 92 per cent as against 83 per cent for men (Hulme 1991). In Malaysia, Projek Ikhtiar reported 95 per cent repayment rates for women as against 72 per cent for men (Gibbons and Kasim 1991).

Our work reveals that in practice lending to women is much more complex than this assumption chain posits. The following discussion focuses on the situation in Bangladesh, where such an emphasis has been best developed and where a number of other studies looking in detail at the empowerment impacts of credit have been conducted.

Turning first to isolation, there is a strong case that credit programmes can reduce the relative isolation of women. Anyone who has witnessed the village-level meetings of BRAC and Grameen Bank (and other agencies) cannot but be impressed by the bringing together of large groups of women. In these regular meetings women conduct their savings and loans activities and have an opportunity to share information and discuss ideas. Such opportunities previously did not exist. Measuring the results of such association was beyond the limits of this study but the creation of a regular forum at which large numbers of poor women can meet and talk represents a 'breakthrough' in the social norms of rural Bangladesh.

In terms of empowerment (that is, reducing 'powerlessness') our study reveals the naivety of the belief that every loan made to a woman contributes to the strengthening of the economic and social position of women. The evidence on this issue is mixed and contradictory. Goetz and Gupta's (1994) work indicates that some 39 per cent of the loans provided to women by four agencies (Grameen Bank, BRAC, Proshika and BRDB) are either fully or significantly controlled by men, while a further 24 per cent are partially controlled by men (that is, men control the productive process while women provide substantial labour inputs). When women do fully or significantly control loans, this is most commonly associated with 'traditional' women's activities (particularly livestock and poultry) so it represents the reinforcement of existing conceptions of the economic role of women. White (1991) has reported similar observations for other credit schemes in Bangladesh, while studies of TRDEP report that 'wives, sisters or daughters' are often token members of groups and that 'in general TRDEP does not serve women's interests will' (IDSS 1994: 3). Such high levels of male appropriation of female loans led Goetz and Gupta (1994: 1) to conclude that 'gender relations and the household are in effect absorbing the high enforcement costs of lending to men in Bangladesh's rural credit system because women have taken over the task of securing loan repayments from their male relatives'. From this perspective credit schemes are doing relatively little to empower women, though one important exception must be noted. By contrast, the same study reports that 55 per cent of widowed, separated or divorced women fully control their loans (compared with 18 per cent for women in general) and that only 25 per cent of such loans are fully appropriated by men (*Ibid.*: 13). Given that such women are usually regarded as the most vulnerable in Bangladeshi society this suggests a significant advancement in their capacity to engage in economic activity.

Schuler and Hashemi's (1994) study of Grameen Bank and BRAC female members sought to measure their degree of empowerment[15] and relate this to contraceptive usage. Their data on the effects of credit programmes on the position of women presents a very positive picture (Table 2). The proportions of Grameen Bank and BRAC female members who were 'empowered' were at much higher levels than the control group, and the proportion of women 'ever beaten by husband' who had been beaten in the previous year was significantly lower. However, while Grameen Bank membership is associated with statistically significant higher levels of contraceptive use, BRAC's positive impacts are not significant. This difference is explained in terms of the Grameen Bank's greater effectiveness in strengthening women's economic roles and its more disciplined and regimented approach to group activity.

Despite the contradictory nature of the evidence cited above, important conclusions can be drawn. The first is the refutation of the claim that every loan to a woman is a step forward in the empowerment of women. This means that female participation rates in financial schemes cannot be treated as a direct indicator of female empowerment. Although donors are keen to count female scalps (and as a result micro-finance institutions are usually keen to count female scalps), programmes that genuinely seek to empower women will need to research loan usage much more carefully and will need to place a greater emphasis on developing new and more productive economic roles for women. A simple emphasis on disbursing to women is likely to encourage tokenism and the reinforcement of established gender roles (as in the case of loans to women in TRDEP and SACA).

---

[15] In terms of mobility, economic security, ability to make small purchases, ability to make larger purchases, involvement in major household decisions, subjection to violence, political and legal awareness and involvement in protests/campaigns (*Ibid.*: 68).

Table 2: Credit Programmes, Women's Empowerment and Contraceptive use in Bangladesh

Indicator	Grameen Bank Members	BRAC Members	Comparison Group
Mean empowerment score[1]	5.0	64.0	9.0
per cent of sample 'empowered'[1]	27.0	59.0	4.9
per cent beaten in last year by husband	56.0	15.0	35.0
per cent beaten in last year by husband as a share of per cent ever beaten by husband	47.0	3.5	23.0
Currently uses contraception	27.0	46.0	43.0

Source : Schuler and Hashemi (1994)

Note : [1]The indices used to compute this index are listed in endnote 15. For a full definition, see Schuler and Hashemi (1994: 67-70).

Secondly, the exact nature of 'who' women's empowerment is to be judged against needs to be clarified. Our work on BRAC led to the finding that 'those women who do particularly well as a result of credit are much more likely to be empowered *vis-à-vis* other (less well off) women, rather than *vis-à-vis* the men folk in their household (or in wider society)[16]. A similar case could be made against the Grameen Bank whose members have significantly higher educational levels than the wider female population (29 per cent of GB female members have attended school as compared to only 18 per cent of a comparator group: Schuler and Hashemi 1994: 69). So agencies need to pay much greater attention to their capacity to assist target groups within the female population (particularly the assetless, widowed and divorced) rather than treating women as a homogenous group.

Finally, our study indicates the need for further research on the question of whether credit groups are also 'women's groups' (that is, associations of women that seek to promote women's needs and interests outside the field of credit). The accounts (Lovell 1992; Fuglesang and Chandler 1993) of BRAC and Grameen Bank groups acting autonomously and asserting women members' inter-

---

[16] It is no coincidence that BRAC field staff (the vast majority of whom are young males) report that female members are much 'easier to work with' and 'disciplined' than male members. Female members are observing their cultural norm in relation to a male.

ests *vis-à-vis* their menfolk are fascinating but are premised on the erroneous assumption that such incidences did not occur in earlier times. If anything, the scaling-up and professionalisation of BRAC and Grameen Bank has weakened the position of female (and male) *shomitis* as belonging to the credit programme and not as belonging to the members. This situation contrasted markedly with our Sri Lankan case, where SANASA's primary co-operative society members controlled society activities (Montgomery 1996).

The Poor Amongst Poor

While case study institutions have extended the reach of the formal financial sector into 'upper'—and 'middle'-income poor households, as discussed in earlier sections, they have been relatively ineffective in reaching the poorest. Several factors have frustrated such efforts. First and foremost is the emphasis on credit delivery by many institutions: for the poorest people and households the opportunities for credit-financed self-employment are very limited, and the risks are unreasonably high. As Rutherford (1993 and this volume) argues from his research with the very poor in Bangladesh, the poorest commonly practice 'self-exclusion' from income-generating credit initiatives which they do not perceive as a solution to their livelihood problems.

Second, for group schemes, processes of social exclusion are important: that is, group members (most often people below the poverty line) deciding that some prospective members are 'too poor' to be given group membership. This may be on the economic grounds that such folk are 'too risky' or on social grounds as the poor differentiate amongst themselves. Direct cases of such exclusion were reported for SANASA, BRAC, TRDEP and SACA where members identified some people in their villages as being unsuitable for group membership because of the intensity of their poverty.

Third, there is evidence that as credit programmes are expanded and management is professionalised, the incentive structures for staff (bonus payments and promotion prospects) favour a concentration on groups other than the core poor[17]. This was clearest in the case of BRAC (Montgomery, Bhattacharya and Hulme 1996) where

---

[17] Tapered interest rates (charging smaller borrowers high interest rates) may partially offset such tendencies.

the average value of new members' assets was higher than the asset levels of successful third-time borrowers. Field staff find that headquarters-set performance targets can best be achieved by working with the poor rather than the core poor (*Ibid.*). In Sri Lanka, SANASA is currently wrestling with this dilemma as it shifts from a predominantly voluntary staffing basis to a more professional one.

Extending financial services to the poorest will require innovations beyond those developed by the present generation of 'microcredit for the poor' institutions. The starting point for such experiments lies in easy-access savings and small contingency loans.

The Disabled, the Old and the Infirm

None of our case study institutions was observed to provide services for people with significant disabilities[18] during the course of fieldwork. This should not be a surprise, as the emphasis on self-employment in low-demand-high-competition markets for goods and services made many programmes irrelevant in terms of the opportunities available to the disabled. Removing the need for collateral makes little difference to the disabled who face many other obstacles. Interestingly, even BRAC's Income Generation for Vulnerable Group Development (IGVGD) programme—targeted on the core poor—has a physical-fitness criterion for access which ensures that the physically disabled are screened out.

The point of the above observation is not to argue that innovative financial institutions should service the needs of the disabled or infirm (it would be unreasonable to expect them to meet all social welfare needs), rather, it is to point out that explicit and implicit claims that such programmes reach the 'poorest of the poor' need to be tempered. If we update Doyal's (1983) estimate of around 350 million physically and mentally disabled people in developing countries, then, given population growth, the emergence of AIDS and the explosion of cripplings associated with the increasing incidence of warfare (and the cheapness of land-mines), there is clearly a vast army of disabled poor people who are presently beyond the reach of even the most innovative institutions.

---

[18] The definition and assessment of 'disability' are complex tasks. For the purposes of our field research 'significant disability' simply refers to observable physical attributes or behaviour that mean that the mobility and capacities of an individual are clearly well below those considered the norm in that specific setting.

Similar arguments apply to many elderly people. However, the demands of the elderly for accessible savings schemes was evident from Sri Lanka where almost 10 per cent for SANASA members had reached 'retirement age'. Such people used SANASA to store 'lumpy' income, from the sale of assets or gifts from friends and relatives, and to accumulate savings so that major purchases could be made (such as roofing iron and new clothes). The needs for financial services for the elderly may well be similar to those of the able but very poor.

Reconceptualising 'Finance for the Poor' and its Policy Implications

Our main finding is the need for the designers of financial services for poor people to recognise that 'the poor' are not a homogenous group with broadly similar needs. The emphasis of the last ten years has been on a promotional model of poverty-alleviation through micro-credit financed enterprise expansion. Such a model is valid for 'middle' and 'upper' poor households, with members who have entrepreneurial flair. It is inappropriate for the poorest households, those with a high degree of income insecurity and the disabled. Commonly credit for micro-enterprise promotion and finance for poverty-alleviation will overlap, but this does not mean that they are the same activities, as is often assumed by micro-finance institutions and their sponsors.

Recognising the heterogeneity of the poor clearly complicates matters for scheme designers because potentially numerous groups and sub-groups could be distinguished. For analytical purposes it is appropriate, at least as a starting point, to conceptualise two main groups within the poor: the core poor who have not crossed a 'minimum economic threshold' and whose needs are essentially for financial services that are protectional, and those above this threshold who may have a demand for promotional credit. This minimum economic threshold is defined by characteristics such as the existence of a reliable income, freedom from pressing debt, sufficient health to avoid incapacitating illness, freedom from imminent contingencies and sufficient resources (such as savings, non-essential convertible assets and social entitlements) to cope with problems when they arise. The distinction between the poor and core poor must be applied to both women and men. Although females and those within female-headed households comprise a disproportionate number of the poor and core poor in most situations, simply targeting activities on women is too crude a device to help the poorest. The female population has a range of living standards and so poverty-reduction needs to target specific groups carefully. While these vary from place to place they will commonly include the widowed, the divorced, the disabled and the elderly. Recording the participation rates of such people with financial institutions, something which none of our case studies regularly did, would be a useful step in the direction of more effective targeting.

If the poor/core-poor categorisation is accepted, then it indicates the need for either more comprehensive approaches to financial services than were offered by most of our case study institutions (that is, providing investment, consumption and contingency loans and offering high accessibility savings schemes) or for specialist institutions to operate to meet the differing needs of poor households and core poor households. While the financial viability of institutions genuinely targeted on the 'core poor' remains to be tested, the profits generated by our Indonesian cases from savings schemes, by SANASA's primary societies from 'high interest' contingency loans, and by official pawnbrokers (in Sri Lanka and other countries) indicate that there are substantial possibilities for operational recovery.

Those concerned with the poverty-reducing impacts of financial intermediation must note two further points:

1. Even if future innovations lead to the design of financial services that meet the needs of the poorest, the requirement for other social security mechanisms—employment guarantee schemes, food-for-work programmes, drought relief, indigenous welfare practices, the conservation of common property resources—will continue. Financial services targeted on the poor are only one element of a national poverty-reduction strategy, not a replacement for other approaches;

2. For the 'above-threshold' poor the evidence from Bangladesh indicates that many close to the poverty line may be able to invest in activities that can offer increasing returns over time (that is, the TRDEP borrowers). For less well-resourced, less well-educated and less well-connected poor people (the middle poor) then investment opportunities are very limited and are likely to remain so, unless the formal economy experiences substantial growth. As a consequence, if a promotional strategy

is being pursued, this category of micro-entrepreneur is likely to need assistance in the areas of product development, technical change and marketing if income-generating credit is to be anything more than 'one step up' the income pyramid. The risks of technical innovation must, at the very least, be shared between institutions and borrowers, and not simply absorbed by borrowers as in the case of BRAC's deep tubewell programme (van Koppen and Mahmud 1995).

The recognition of the heterogeneity of the poor makes the identification of 'best practice' design features for poverty-reduction difficult. If we pose the question 'Which scheme has most rapidly pushed household incomes above the poverty line and is associated with continuing income growth from successive loans?', then TRDEP and BancoSol are highly successful. However, if we pose the question 'Which schemes have helped to raise and protect the incomes of the poorest households', then we would judge TRDEP and BancoSol failures. The challenge of providing financial services to poor people needs to be seen not as developing a 'super institution' that meets all of the needs of all of the poor: rather, the challenge is to develop a set of institutions (in any one area or country) able to serve the differing needs of the poor. Institutional pluralism and competition (for market share and for subsidies) are required if financial services are to evolve that serve the needs of the poorest: the Grameen example, which currently dominates the field of institutional design (namely, TRDEP, BRAC's RDP, K-REP Juhudi and MMF), must not be seen as the only approach, given this model's clear problems with working with the 'core poor'. Further experimentation with protectionally focused schemes for the poorest, offering savings and contingency loan services, perhaps on an individual basis or on the basis of indigenous savings societies, is needed to explore whether a 'second wave' of innovation can provide services to the poorest. Other chapters in this book address this issue.

A final issue to note is our finding that there has been a gross exaggeration in that part of the literature[19] on credit schemes, claiming that they are vehicles for the mobilisation of poor people

---

[19] IDSS (1994) refers to sections of the literature on the Grameen Bank as ranging from the adulatory to the sycophantic. Lovell's (1992) work on BRAC falls somewhere between these two positions.

that are, or soon will be, stimulating dramatic social and political change. Innovative financial services, as was illustrated in the discussion of gender relations and of relations amongst the poor, have complex socio-political impacts. Commonly, they simultaneously change some dimensions of social relations (such as reducing the isolation of women) whilst reinforcing and strengthening other relations (such as a gendered division of labour). The financial interventions studied in this volume are only one of the many forces for change operating in the countries concerned and none of our case studies have created forms of class or interest group solidarity of structural significance to date. The largest institution, the Grameen Bank, has moved away from its earlier focus on 'mobilising the poor' to being a specialist bank, while BRAC has effectively abandoned its village-level institution-building activities and is no longer attempting to federate village organisations. SANASA makes a local-level contribution to notions of democracy in Sri Lanka but it explicitly avoids a political identity. The Indonesian institutions work with individuals and make no attempt to establish new social organisations. SACA was an extension of the Malawian state and, if its members had a group position, this was conservative and orientated to the status quo. Financial institutions—even those that mobilise large memberships—can help the poor in terms of their economic impacts and influences on micro-level social change. They are not, however, vehicles for social mobilisation that will confront existing socio-political structures (Hulme and Mosley 1996: 138-156).

Conclusion

Recent innovations in financial intermediation have permitted a number of institutions to make a contribution to poverty-reduction in terms of raising the incomes of some poor people and (in a smaller number of cases) helping to reduce income vulnerability. Given this evidence, there is a strong case for their extension into new areas, allied to a recognition that such schemes are not a panacea for poverty-reduction.

Most contemporary schemes are less effective than they might be because (a) they treat 'the poor' as an undifferentiated group, and (b) they focus largely on a promotional strategy for poverty-reduction involving a rigid loan disbursement regime rather than more diverse credit and savings services. As a consequence, the poorest people have little access to these schemes, are likely to take on unreasonable risks if they do participate, and the benefits are most likely to accrue to 'middle'—and 'upper'-income poor who have crossed an economic threshold that means a major part of their income is secure. A reconceptualisation of financial services for the poor—that recognises both promotional and protectional strategies and (at a minimum) differentiates between the poor and the core poor—is now necessary. As a result a further phase of institutional experimentation and innovation—a second wave—is required to extend financial services deeper down the socio-economic pyramid. Ironically, it is the success for the 'first-wave' finance-for-the-poor schemes, and particularly the Grameen Bank, that is the greatest obstacle to future experimentation. Most designers and sponsors of new initiatives have abandoned innovation, and 'replication' is leading to a growing uniformity in financial intermediation for the poor.

Appendix 1

INSTITUTIONS STUDIED

BRAC-RDP	Bangladesh Rural Advancement Committee-Rural Development Programme
BancoSol	BancoSol, Bolivia
BRI	Bank Rakyat Indonesia Unit Desas
BKK	Badan Kredit Kecamatan, Indonesia
Grameen	Grameen Bank, Bangladesh
KIE-ISP	Kenya Industrial Estates—Informal Sector Programme
KREP	Kenya Rural Enterprise Programme
KURK	Kredit Usaha Rakyat Kecil, Indonesia
MMF	Malawi Mudzi Fund
RRBs	Regional Rural Banks, India
SACA	Smallholder Agricultural Credit Administration, Malawi
SANASA	Thrift and Credit Co-operatives, Sri Lanka
TRDEP	Thana Resource Development and Employment Programme, Bangladesh

Chapter 4

THE POLITICAL ECONOMY OF MICRO-CREDIT

Rehman Sobhan

Twenty-five years have now passed since the first micro-credit programmes were introduced in Bangladesh, with the Comilla initiatives of the 1960s and later, those of BRAC in Sylhet. In fact the silver jubilee of micro-credit in Bangladesh could well be celebrated at the same time as the liberation of Bangladesh. The conference from which the papers in this volume were drawn could be described as a taking stock of the accumulated experience of the past quarter of a century, as well as a glance ahead at the next twenty five years: an assessment of what has passed and plans for what will be. I think this is very timely.

I would like to offer here a number of thoughts which have come to me as I have reflected on micro-credit from the position of one who has casually interfaced with it in an academic and occasionally in a policy-making context. In the following paper I will propose ten issues which I think will be germane to the future of micro-credit in Bangladesh and indeed in other developing countries.

The first and the most obvious is that micro-credit originated with poverty alleviation as its objective, with social development or mobilization as it is now known being added to the agenda later. In terms of poverty alleviation, the relevant question which empirical research has attempted to address has been the capacity of micro-credit programmes to actually solve the problem. Here I think it is very important to distinguish between the micro analysis and the macro impact of micro-credit. Micro analysis consists of reviewing various poverty alleviation programmes. A recent major study has been the World Bank/ BIDS study which looked at the impact of three major institutions: the Grameen Bank, BRAC and BRDB as delivery agents. Such studies appear to have established that credit programmes which have delivered resources to the poor have, at the level of household income and expenditure, had some impact on improving the living conditions of their particular beneficiaries. Broadly speaking, that seems to be the conclusion which has emerged from most of the studies.

However, the conference and the contributors to this volume raise the important question of graduation. While there is an improvement in the conditions of life of those beneficiaries of micro-credit, we must question the sustainability of the graduation of these groups— might there be at some later stage a regression? Perhaps at the moment research has not reached the stage where it can give a definitive answer to the question. I am interested to note that this question is being considered although contributors appear not to have come to any definite conclusion about the capacity for micro-credit programmes to actually bring about a sustainable transformation in the lives of the beneficiaries, such as would allow them to move into a different social category as a consequence of their access to micro-credit. Is this outcome a limitation of the research or a result of the limitations of the programme itself? This is an important issue which really deserves to be addressed.

Another issue which has been highlighted in Bangladesh is the limited impact of the micro-credit programmes on the very poor and here again I do not know if the research initiatives in this area have developed the right analytical framework. Is there in fact a very serious problem, which is that micro-credit programmes tend to be made up of groups of people who are most likely to be able to repay what they borrow, who are therefore less of a credit risk to the group as a whole, and that this type of person tends to form the constituency of the micro-credit programme, to the exclusion of the more needy? Again, I am not quite sure whether the conclusions on this are sufficiently robust to justify the consideration of new sets of interventions in regard to micro-credit.

Moving on from here I think you come to a much more serious problem, which is the whole issue of the relationship between twenty-five years of micro-credit and its macro impact on such issues as poverty alleviation and growth. I recall putting together our independent review of Bangladesh's development, the report which is prepared every year by the Centre for Policy Dialogue (CDP). In our chapter on poverty, we address the theme of the role of NGOs. One of the interesting points which came out of this was that, at the macro level, there has been an enormous impact from the provision of credit (originating of course largely from the Grameen Bank but certainly followed by the NGOs). The numbers, when you put them together, indicate that these institutions have in fact become the largest source of institutional credit in the Bangladeshi rural economy. This puts them as credit deliverers in a class with the rural lending programmes of the commercial banks. In fact we are talking about something like two-thirds of institutional credit now originating from the Grameen Bank and the NGOs. This is a not insignificant set of figures, and when you are talking about numbers of this size, then you must obviously be looking for the macro impact at this level of new lending programmes which were not in existence twenty-five years ago. You would expect that this should have had some transformatory effect on the macro scene both in terms of poverty alleviation and growth. Now, here again I find that the level of research and analysis has not satisfactorily addressed the relationship of micro interventions on the macro economy. Perhaps this is partly a methodological problem—the models of analysis that are available have really not performed this task satisfactorily. At the moment there is a very serious gap in the literature. At the same time this poses a very serious problem for the people who are involved in micro-credit. While every micro-credit agency should be complimented on the fact that they have improved the conditions of life for a substantial number of individual borrowers, most of them have come into the whole process of offering micro-credit as part of a wider agenda of poverty alleviation, if not actually poverty eradication. The whole theoretical hypothesis behind this was that, if you provided credit on a sufficient scale this would both contribute to opening up the areas of opportunity for the poor which would have an impact on production relations in the rural economy, and would actually have a social transformatory effect. This ties in with the much broader question of credit and social mobilization.

Recollecting again the findings of the BIDS/ World Bank study, the various village studies seem to have established two things. First, that it is evident that at the household level there has been an improvement—though there seems to have been some variation between the different credit programmes—but second, and more seriously, at the macro village level, there was not a very significant dent made in the levels of poverty. So, even after twenty or twenty-five years of credit, in villages where micro-credit had been playing a very significant part, reported levels of poverty were still in the range of anything from 50% to 60%—certainly above the national average. It may well be that the villages which were originally selected when credit groups or institutions were setting up their programmes were in fact poorer than average, but these figures would indicate that, (disregarding the original parameters in the choice of clients or the villages where clients were located) micro-credit programmes have not made a very significant breakthrough in the rural economy. Have these micro-credit programmes then created a degree of critical mass which would bring about an element of structural transformation and sustainable growth in the village economies where micro-interventions have actually been located? One would assume that with twenty-five years of micro-credit going into a particular area certain transformatory effects on the macro-economy should have been felt. I am not merely referring to the village economy, a macro entity in itself, but also to the national economy, where the poverty alleviating impact at a macro level of credit interventions and the social transformatory effect of this particular process should be felt. I have no satisfactory answers for this but I am sure that those who are reflecting on this experience and who are engaged at the operational level in alleviating poverty where micro-credit has become an important instrument would certainly be giving this matter some thought. It is certainly important to try to discover from the literature on offer as well as from field experience, why micro interventions have had no discernible macro economic impact on poverty; on broader economic growth; and in wider structural transformation of the political economy.

Moving on from this, I would like to outline some of the problems which micro-credit interventions have faced over the course of the last twenty-five years. My own hunch is that the whole micro-credit experience has inadequately addressed the issues of credit in relation to the market with its peculiar imperfections and structures, and credit in relation to the actual substantive production structures of the economy—i.e., those elements of the economy which are substantially contributing to its growth, the whole issue of social structures and production relations. Presumably none of those involved in micro-credit went into it thinking of themselves as grassroots bankers, most of them went into it thinking that they were in the process of transforming the broader conditions of life of the poor and the situation of the poor in rural society. The nature of micro-credit seems to have been that you give credit to people and you leave them to fend for themselves within the existing structures of the market. Perhaps there are more innovative efforts of which I am unaware, which attempt: to relate the credit interventions to an assessment of the dynamics of the markets where credit is in fact on offer; to service their clients to see whether this is stimulating activity on the part of the borrowers in the more dynamic components of the market; or experimentally to just leave them to operate at the level or skills they have at their points of entry into the market, (i.e., at their existing capacities).

Clearly there is a whole range of activities going on in the market which determine the basic structures of the rural economy and rural society. Here we are obviously thinking in terms of the main sources of production and exchange; the agricultural sector; the more dynamic components of rural small industry; and some of the major trading initiatives which take place (trading in food grains, agricultural inputs, commodities being produced and consumed in the rural economy). These all determine the relationships between the different social and economic groups which operate. Unless the rural poor are able to access the major earning opportunities which exist in the rural economy there is unlikely to be a very significant impact on the sustainable alleviation of poverty; thus micro intervention will not have a major impact on the broader macro society and macro economy. How satisfactorily have the micro interventions addressed this problem? Here I am referring to the sort of areas which you see in quite a number of other countries, where micro-credit to the poor has been used to get them into, for example, sustainable grain trading—the whole issue of strengthening the poor so that they are not forced to sell when they are most vulnerable but can wait to sell their products when the market is more favourable. These are issues which have concerned the rural economy and the conditions of the poor not just for twenty-five years but from time immemorial—how does micro-credit help the small farmer or even the subsistence farmer who at harvest time is compelled to sell some of his output in a very unfavourable market environment? What do you do about the poor jute grower who, again from time immemorial it is supposed, has been victimized by the imperfections of the market and the unequal relationships between the grower and the traders? I think this must bring one to conclude that the credit process does not really address the key elements involved in sustaining the rural economy. It has not really had any significant impact on agriculture. I have never understood why agricultural credit has remained the monopoly of the *mohajan* or the informal sources of agricultural credit. Certainly the Krishi Bank has not made any inroads at all into the traditional sources of borrowing which still remain the main source for stimulating production and employment in the rural economy albeit at usurious rates, thus depressing potential activities. In considering the agricultural sector there must be an awareness of its trading opportunities and of the value-addition components in the agricultural economy, prompting the key question: why has there been no breakthrough in the micro-credit interventions in touching on the substantive elements of the rural economy?

I am therefore concerned about the whole issue of production relations. I am aware that, in this post-socialist era, discourse on production relations has become somewhat unfashionable, although it is still discussed—using less inflammatory phraseology perhaps—but in the end it comes down to who owns the land, who owns the productive assets, and how this contributes to empowerment within the social relations of rural society. Once again this issue has existed not for a mere twenty-five years, but from time immemorial, with attempts to rescue the poor agriculturists, the rural poor from the hegemonic social influences of the money lender, the landlord, and various other people who use to their advantage the inequitable access to key agents of production—land, capital and of course skills, and more recently in the aid-driven system, access to aid projects, aid resources and various other forms of empowerment which have come through the delivery of aid resources into the rural economy. These relationships are crucial and at the end of the day if you transform production relations in terms of access to these key assets, you bring about a process of social transformation wherein the poor in the village can enjoy a new position in village society. How far has micro-credit really addressed these issues of changing aspects of production relations, access to land and other assets? I know that Proshika has addressed this to some extent by attempting to make the landless into water-lords. But has this had a sufficiently transformatory effect on the rural economy where there is now a substantial class of water-lords, largely those in command of the shallow tube wells but including some who have gained possession, at virtually no cost, of the deep tube wells and who have then, purely to improve market conditions for themselves, closed down most of the deep tube wells? What in fact have micro-credit interventions really done to bring about a social structural shift in production relations? If there has not been a shift, is it beyond the conception or beyond the capacity of those intervening through micro-credit, I wonder? Obviously this too has a wider political manifestation, which I will touch on when I go on to talk about credit and social mobilization.

Contributors to this collection address the issue of financial services—I notice that we now refer to the more up-market financial service conceptualization of "the delivery of credit" where once we would have talked of "helping the poor through micro-credit". I am pleased to see that our terminology is keeping up with the fashions, but at the end of the day the issue is that there is a real problem of relating to the way in which financial markets work. Hashemi points out the large accumulation of savings in the Grameen Bank and I presume that if you add up the savings assets of the NGOs in aggregate this should also add up to quite a substantial sum. Why has this not had a structural impact on the way in which financial instruments have been fashioned by the financial markets? Why have financial instruments even at the micro level not really emerged to access this—I mean why is it that the debate about savings is restricted to localized, micro use, such as allowing a man to pay for the hospitalization of his wife if she falls sick? These savings assets are now large-scale resources and presumably if you were thinking in terms of efficiently functioning (I am not talking about perfectly functioning, but at least a profit-motivated financial system) you would be developing financial instruments to deploy these accumulated resources to secure their growth. Why has no one done this, why have the mutual fund operators in the Dhaka Stock Exchange not made any attempt to develop instruments to reach out to this? I remember bringing this to the notice some of the big brokers in the stock markets and asking them why they have not started talking to the Grameen Bank and other agents about accessing these resources and putting these financial assets at their disposal? Perhaps they are incompetent, or unaware that these constituencies exist, but either way, why have the micro-credit operators not appreciated the potential offered to them in deploying such resources by becoming players in the institutionalized financial market even though operates in a highly imperfect and inadequate way? This is clearly another issue which must be addressed. You have to develop instruments which do not exist, you have to develop a willingness, a boldness, a capacity to come into markets which do exist. This does not seem to have happened to any very significant extent as far as I am aware.

I now turn to the issue of insecurity and regression in the ranks of the poor due to their exposure to exogenous shocks ranging from floods and illnesses to simply the traditional insecurities of life. The key question here is whether access to credit designed to improve the cash flow of the rural poor, can now be given more formal and positive acknowledgement—rather than just talking of its productive use while somehow thinking that consumption credit is itself undesirable? What conclusions can be drawn from this—savings designed to help people out have certainly been generated, but do the current institutional interventions deal adequately with the issue of insurance? What major efforts are being made to use or develop financial instruments in the insurance market which will substitute for more traditional but waning forms of security? If financial institutions in the society are not providing effective insurance at the political and social level against the traditional insecurities which are offered to more or less effective degree (either via the market or the state) in more advanced societies and in quite a number of developing economies, what are the NGOs effectively doing about developing social insurance systems which would protect their poor clients from such insecurities? The resource base now exists to attempt this. The NGOs have the institutional capabilities to set up insurance systems, and by operating collectively it should be possible to move into group insurance initiatives, pioneered by the NGOs or operating with some of the established insurance companies. I believe that Green Delta Insurance and various others have started some modest initiatives. Elements of synergy between the NGO sector and the insurance market should be emerging if it is accepted as a serious problem. NGOs do appear now to have both the awareness as well as the resource base to do something about it.

To move on very briefly to the issue of credit and social mobilization, or social development as it is now called to keep up with the fashions. Obviously the UK ODA would be much happier with the idea of social development rather than the more aggressive concept of social mobilization. Now in Bangladesh, if you go back to the writings of Malcolm Darling in "The Punjab Peasant in Prosperity and Debt" and the old bureaucratic counterparts who were wanting to rescue the poor cultivator from the excesses of the local money lender and the landlord, you are really talking about social transformation, which underlay both humanitarian concerns and subsequently political and social concerns. People like Hasanul Haque Inu (in the audience today) and his counterparts on the Left, have certainly included this in their political discourse from the time that he went into student politics if not before, and their predecessors from the *Tebhaga* movement and various other movements were concerned about the process of social transformation. Much of this discourse was built around the emancipation of the rural poor from the exploitative control of those who were controlling credit markets. So now that NGOs have in fact become major players in the credit market, how far has this had a social transformatory impact? Has there been a fundamental change, a social transformation in relations with the traditionally powerful in the village economy? Does the NGO literature and micro research address this adequately? Certainly I cannot find any conspicuous impact. Looking at the voting patterns at Union *Parishad* level, for example, I could not say that, with some notable exceptions, there has been a significant change in the social composition of those who are holding power at the local level. If these groups have been rescued from the people on whom they were traditionally dependent for their main sources of credit and livelihood, has this been sufficient to emancipate them from these traditional areas of control over their life? If not, is this not a problem of inadequacy? When NGO household coverage goes from 25% to 50% or 75%, will it change? Or is it because NGOs are too thinly spread, because everyone feels it to be a part of their mandate to cover all of Bangladesh, in order to access resources from multiple donors? Or have there been attempts to accumulate critical mass at individual village levels? Has one big NGO gone into one village and said: "We are not going to leave this village until we bring about a social revolution in this village by changing some of the most fundamental exploitative social relations which actually exist here."?

What is the problem? Is it the nature of the way NGOs do their work or is it a problem of human inadequacy? These are questions not for me to answer but for those involved to ponder.

This brings me to the two concluding points which need to be addressed in the future; that of institutional sustainability and the forthcoming the Micro-credit Summit. Institutional sustainability is in its way crucial and needs to be dealt with substantively because it is a problem which ties in with most of other issues which I have already raised. At the end of the day, the notion of institutional sustainability is that you create indigenous institutions which can mobilize resources, lend them, recover them and treat this as a locally underwritten and financially sustainable enterprise. The Grameen Bank claims that it has moved to that level, but how many of the other institutions have actually done so? I do not really know. But presumably after twenty-five years a level of institutional graduation should have been achieved which would be the precondition for institutional sustainability, where the relevant lending institutions will have either generated enough of an indigenous resource capability so that they no longer need to go to external sponsors but can operate within the existing financial markets, or they would have set up micro institutions which are themselves financially sustainable at the village level. Has this happened? I cannot say for sure but my guess is that very little of that has happened and if you in fact look at the generalized thrust of a lot of the poverty alleviation aided programmes, a large part of this is really underwritten by credit delivery programmes, whether through the BRDB or through the NGOs, where enormous, rising volumes of resources are being pumped into the economy through these micro delivery systems. Now if this trend continues and increases for the foreseeable future, instead of a trend towards gradual withdrawal (because NGOs have created sustainable institutions), then I think NGOs have a very serious problem on their hands which had better be addressed. My one hope is that when my son (also in the audience today) is old and grey he too will not be discussing issues of institutional sustainability at a similar conference!

Here again I must address the last point, the Summit. I think what I would like to see from this Summit is thought given to the high cost of delivering aid from the rich country tax payers which underwrite both the international agencies and the bilateral donors. At the moment what I think has happened is that a variety of institutions have been created which deliver tax payers' resources or financial aid into the micro economy, via NGO and Grameen Bank micro-credit programmes. Now if you look at it from a global macro perspective, and I hope that the Summit will do so, these are very high cost and very inefficient delivery mechanisms, starting from the multilateral agencies, going on to the bilateral agencies and the various other mechanisms which deliver resources. If you are essentially delivering or aiming to deliver a thousand *takas* to a poor village woman in Dinajpur (from the tax paid in US dollars by a woman in Minnesota which is collected by the US government treasury department), what does it actually cost to deliver that one thousand *takas*? My hunch would be that it would probably cost about five hundred *takas* to make that delivery. Perhaps, as a suggestion, what is needed is a revolution in the global postal system whereby the lady in Minnesota can directly send one thousand *takas* to the poor woman in Dinajpur, which may be perhaps the most cost-effective way of delivering aid to the poor. But in the meantime these institutions have become part of the problem and they have led to what I like to call "system loss" in the aid system. The system is eating up half the money which is designed to address the problems of poverty alleviation from the big institutions. The consultancy work, not to mention the enormous amount of academic research, these are all costs and they come out of the pocket of the *saree* or the bag of the poor woman in Dinajpur. As an amateur these thoughts have occasionally come into my mind. I presume the NGOs as professionals living with this have also thought about it, but since they remained unanswered questions I presume that on this twenty-fifth anniversary there is a real desire to start addressing them so that similar exploratory discourses with inadequate answers twenty-five years from now can be avoided. I congratulate the NGOs for getting the debate going and I hope that some constructive ideas will have emerged from this conference which will impact on the way in which they operate.

PART II

CASE STUDIES AND THEMES

Chapter 5

EXPERIENCES AND CHALLENGES IN CREDIT AND POVERTY ALLEVIATION PROGRAMS IN BANGLADESH: THE CASE OF PROSHIKA

Yuwa Hedrick-Wong, Bosse Kramsjo, Asgar Ali Sabri

Overview

The standard analysis of the failure of the formal credit system in serving the rural poor stresses two aspects of "market failure". First, there is a perceived situation of imperfect information. This refers to poor knowledge of, on the part of the formal credit system, the business environment in the rural sector, and in particular a profound ignorance of how such credit could be used to develop a viable investment with healthy returns. Thus, the result is a perception of great risk in lending to the rural poor. Secondly, there is a perceived situation of imperfect enforcement. Due to the rural poor's lack of collateral, loans given to them are therefore seen as very difficult to recover should defaults occur. Jointly, these situations of imperfect information and imperfect enforcement have made lending to the rural poor appear extremely difficult if not outright impossible.

Attempts to provide credit services to the rural poor have to overcome these difficulties if the lending institution is to stay financially viable. And it is now a generally accepted analysis that social services are required to accompany credit services if these difficulties are to be overcome. Virtually all organizations and agencies that have attempted to overcome the difficulties described above provide organizational and social development assistance to the poor. They typically offer group-based lending with strong assistance in the organization and training of groups. Strict observance of the norm of group behaviour encourages members to be socially and economically accountable to each other, thus creating pressure among group members to monitor and enforce the contracts.

A recent survey of 13 rural credit programs shows that all share common features in the provision of social services along with financial services.[1] It is significant to note that there is a wide range of service configuration among the different existing models of providing credit services to the rural poor, representing different assumptions regarding how much social development service is optimal. This points to a question that is of paramount importance in understanding the connections between credit and poverty alleviation in Bangladesh: what is the optimal level of social development services that is required to accompany the provision of credit services? The Proshika experience represents perhaps one generic approach—it believes that social and credit services need to be integrated into a coherent development approach to generate sustainable impacts in poverty alleviation.

Proshika Development Strategy and Experience

Proshika is one of the largest national non-governmental development organizations in Bangladesh. Since its inception in 1976 it has been endeavouring to engender a participatory process of development and succeeded in pioneering an approach that puts human development at the centre of its vision.

Proshika's development philosophy can be summed up as "development through the *empowerment* of the poor". In concrete terms, empowerment is achieved through raising the consciousness of the poor, equipping them with organizational and practical skills, supporting them with needed resources, infusing them with the confidence and the determination necessary for taking actions to improve both their social and economic lives.

In practice, Proshika has innovated a process of group formation and training among landless labourers, peasants, rural workers of different trades, and women from the households of these socio-economic groups. Through both formal and non-formal training, these

---

[1] See Alam, Zahirul (1995). "Comparative Study on Conditions and Interest Rates of Rural Credit Schemes in Bangladesh". Rural Poor Program Task Force of BRDB, Report No. 5.

groups learn practical as well as human development skills. While the former helps in improving group members in various income earning skills and trades, the latter addresses something more fundamental-changing the attitude and the consciousness of the rural poor.

Concurrent with group organization and training, the poor are also encouraged to pool their resources to pursue employment and income generating projects. Proshika has instituted a system of "Revolving Loan Funds" to provide needed initial investment[2] to those groups deemed ready to undertake these projects. In addition to initial investment capital, Proshika also provides the necessary technical support to help make these projects succeed.

It is important to note that while these employment and income generating projects help improve the material well-being of the groups, they are intended to have a second, and perhaps more significant impact. Being organized, group members are now linked up with an extensive grassroots network involved in development. This alters their positions in the community vis-à-vis the landowners and local authorities. Their new skills and access to investment capital also help to change their status in the community. Gradually, the outlook and attitude of the poor are expected to change from hopelessness and helplessness to self-confidence and self-reliance.

As the groups mature and succeed in their income and employment generating projects, their dependence on Proshika's Revolving Loan Funds is reduced. Eventually, some of them may become capable of undertaking self-financed projects. These groups remain, however, actively involved in Proshika's overall organizational structure and planning. Their initiatives and inputs in turn drive Proshika's development of new projects and help move the whole organization in new directions. It is in this sense, ultimately, that Proshika's development philosophy of the empowerment of the poor is realized through its development strategy.

The question remains, however, as to whether such an integrated social and economic development approach is necessary, and whether the level of social development service provided by Proshika is optimal. In attempting to answer these questions, the remainder of this paper will:

1. describe Proshika's credit programs and summarize empirical evidence of Proshika's impacts to date,

---

[2] With up to 25% contribution from savings from group members.

ii. identify two criteria for assessing the "optimal" level of social development services required to accompany credit services,
iii. present several case studies illustrating different possible outcomes based on different combinations of social and credit services, and
iv. conceptualize the challenges facing practitioners and researchers in this field.

Proshika's Credit Service

The Credit program of Proshika is implemented through a combination of four services:
i. mobilization and utilization of group savings,
ii. provision of matching credit from a Revolving Loan Fund,
iii. technical assistance through provision of skill and management development training and on-the-ground technical advice and support by technically competent development workers, and
iv. provision of marketing assistance where needed.

This combination of services allows Proshika to pay more attention to activities which develop existing resources as well as creating new resources. The design of Proshika's credit programme can be said to be quite deliberate in aiming to create a basis for sustainable development.

Empirical Evidence of Proshika's Impacts

Given Proshika's development strategy and practice, its impact can be defined in terms of two generic types. The first is "economic empowerment impact", the second "social empowerment impact". Conceptually, economic empowerment impact can be measured by the set of indicators that reflects those empowerment processes that lead to improved income and employment conditions, reduced indebtedness, and greater control by project recipients over conditions affecting their long-term economic well-being.

Social empowerment impact, in contrast, is to be measured by the set of indicators that reflects those empowerment processes that lead to improved social status and power of project recipients in the community (especially from the perspective of women), particularly their increased abilities for organizing and mobilizing themselves to initiate and implement desirable social changes, and in dealing with landlords and local authorities.

The intended outcome, empowerment of the poor, is meant to come about as a result in successes in *both* economic and social empowerment.

It is important to point out that the concept of empowerment impact has a subtle but significant difference from the usual meaning of the term impact. Take economic empowerment impact for example. Economic empowerment impact results when the targeted group has acquired the abilities (technical, financial, organizational, attitudinal etc.) to positively affect their economic well-being. In this sense economic empowerment impact is quite different from, say, income impact. While the latter focuses on what effects the project has had on the actual income of the targeted group at a given time, the former assesses the group's evolving capacities to change positively its *actual as well as potential* incomes.[3]

Similarly, social empowerment impact looks at changes in the targeted group's abilities, over the long-term, to change and improve its social status, mobility and life chances *vis-à-vis* the traditional power structure and limiting belief systems.

Empowerment impact is also different from the often used measurement of rate of credit repayment. From the empowerment perspective, a high rate of credit repayment by the poor is not identical to successful empowerment. The question left unanswered by an indicator like credit repayment rate is what do the poor do to repay the credit? If they have to take out another loan, or sell their assets, or reduce their food intake in order to repay their loans; then a high rate of credit repayment does not reflect successful empowerment, but its very opposite.

To operationalize these two definitions of economic empowerment impact and social empowerment impact, a number of issues have to be taken into account. The indicators to be observed and measured must be specified within the context of project activities to ensure a close fit between these indicators and what they are sup-

---

[3] Taking this distinction to extreme, it is conceivable that economic empowerment impact can be said to have occurred even in the absence of any observed income increase as long as it can be ascertained that the targeted group has acquired the necessary abilities to improve, potentially, its income. The limitation of assessing actual income increase is due to the fact that, for most projects only the necessary conditions are provided by the projects. Exogenous factors (sufficient conditions) at the macro/national level often overwhelm what little difference a particular project has made.

posed to reflect (i.e. which aspects of empowerment). Secondly, the availability and quality of data may constrain which of the specified indicators could be utilized. In some cases, proxy indicators have to be used when direct measurement of such indicators is not available.

Given the considerations above, a total of thirteen empowerment indicators have been developed and operationalized for the purposes of impact assessment. Eight of these are "social empowerment indicators" and five are "economic empowerment indicators". They are:

Social Empowerment Indicators

1. Literacy
2. Health Education and Awareness
3. Family Planning
4. Infant Mortality
5. Empowerment of Women
6. Environmental Awareness and Practice
7. Access to Public Resources
8. Participation in Local Institutions

Economic Empowerment Indicators

1. Asset and Indebtedness
2. Income
3. Savings
4. Investment
5. Market Mobility and Power

From 1993 to 1995 two internal monitoring surveys and one impact assessment survey have been implemented to assess Proshika's empowerment impacts. The internal monitoring survey tracks changes observed in Proshika members' performance as shown in their social and economic empowerment indicators. The impact assessment survey compares Proshika members with non-Proshika members, also utilizing the thirteen empowerment indicators[4].

---

[4] For details of the impact assessment results and methodologies employed, please see "Empowerment of the Poor: Impact Assessment of Proshika Kendra's Experience" forthcoming in 1996.

These surveys show conclusively that Proshika has created significant positive impacts among its beneficiaries. In terms of social empowerment, Proshika members' literacy rates are higher than non-Proshika members, however they are measured. In the health field Proshika households have been shown to be better trained in health education and can afford better health care, with better latrines generally. A higher proportion of Proshika women use contraceptives, and on average they have fewer children per women, and much lower infant mortality rate. A much higher proportion of Proshika households than non-Proshika households have dowryless marriages, and the incidence of physical abuse is lower among Proshika women than non-Proshika women. The proportion of Proshika women who report involvement in income generation activities and having decision making power in all major aspects of family lives is significantly higher than their non-Proshika counterparts. Proshika members have also demonstrated greater abilities to access *khas* land and ponds for productive purposes.

From the perspective of economic empowerment, Proshika households demonstrate higher income (27% higher), higher savings (239% higher), more investment (40% higher), and have more assets (57% higher) than their non-Proshika counterparts. The overall rate of return of Proshika households on their investment is about 25% higher than non-Proshika households. When the data are desegregated by genders, the level of achievement of Proshika women's economic empowerment appears to be much higher than Proshika men's.

The question that needs to be answered in this context is whether the level of economic empowerment achieved by Proshika members could be achieved without social empowerment assistance and services, or with the latter provided at a level of lesser intensity. In other words, is the level of social empowerment assistance provided by Proshika "optimal" for the level of economic empowerment achieved? The surveys cited above could provide only some very partial indications.

A few correlation analyses have been carried out utilizing the survey data.[5] Income is shown to be positively correlated with education, particularly among the 6 to 18 years old. Income is also shown to be positively correlated with members' organizational capacity in carrying out projects in social forestry. Furthermore, results of corre-

---

[5] Not all data collected in the surveys are amenable to correlation analysis. Only continuous variables are suitable, thus limiting the number of correlation analyses that could be performed.

lation analysis also point indirectly to Proshika members being able to obtain a higher return on their investment because of their group solidarity and social awareness. These results, however, are too patchy and insufficient to settle the question posed above. One alternative way to address this question is to try to establish some appropriate criteria for assessing what would constitute an optimal level of social development/empowerment service.

Criteria for Assessing the Level of Optimal Social Development Services

Optimality is typically defined as a level of operations where the last additional unit of input is still producing an increase in output, where the very next additional unit of input does not do so. In other words, it is a level of operation just short of decreasing marginal returns. In the case of a factory, the level of optimal operations is at the point where maximum outputs could be realized with the least amount of inputs. In the case of a Jumbo jet flight plan, it is the longest and safest route possible with a minimum amount of fuel and time. In the case of a community, it is a highest satisfaction attainable by residents with the fewest and least amount of resources and services. From the point of view of providing social development/empowerment services, and following this definition of optimality strictly, an optimal level of input would suggest that anything less than such a level would miss opportunities of producing additional benefits for the poor.

While analytically rigorous, such a definition is difficult to apply in a large category of situations such as the one attempted here. In such situations, an acceptable alternative is to specify the minimum conditions for optimality. This allows a way of operationalizing the definition within the concrete context of the situation at hand, while sacrificing some of the analytical rigor. If a set of criteria could be identified as constituting the minimum conditions for optimality, then it could be suggested that when these criteria are met, the level of optimality is being approximated.[6]

Two possible candidates for serving as criteria for assessing the optimal level of social development/empowerment service could be:

    i. whether the development services could reach the "hard core" poor; and

---

[6] This is, of course, in no way suggesting that the optimum is achieved when these criteria are met.

ii. whether the economic development achieved is sustainable once the assistance is withdrawn.

If the hard core poor are not reached because of an insufficient level of social development/empowerment service, then the situation is clearly sub-optimal as the development efforts fail to affect the target group—it is well known that it takes far more social development/empowerment assistance and resources to work with the hard core poor. To the extent that economic development results are not sustainable, the situation is again sub-optimal because genuine empowerment has not been achieved.

There may well be other criteria for assessing what could be construed as the optimal level of social development/empowerment. The two criteria described above could be seen as the minimal conditions required for establishing the optimum which can be demonstrated by the following case studies.

Case Studies

To illustrate the relevance of the two criteria described above, four case studies are presented below. The first can be construed as a case of successful sustainable development. The second and third case studies can be seen as illustrating the importance of achieving both social and economic empowerment in order for poverty alleviation efforts to be successful. They also point to the need for social development services in order to reach the hard core poor. The fourth case can be seen as an example of how economic empowerment alone is insufficient without adequate social empowerment.

Case Study 1: Proshika "Self-Reliant" Groups

In the spring of 1995, a number of Proshika A+ groups (the best performing groups are defined by Proshika's group categorization system) were visited by Proshika's impact assessment consultant in the Bogra region, and group members were asked explicitly their visions of their future, and how they see their relationships with Proshika over the long run. Almost without exception group members mentioned that they would like to go into bigger investment projects with their own savings. A significant number of respondents (over 50%) suggested that they would not need further loans from Proshika. An overwhelming majority stated that they would need technical assistance from Proshika in the forms of small business planning, market analysis, accounting/book keeping etc. but would be prepared to pay for such services should Proshika be able to provide them. They all stated in the strongest terms that they would have to maintain their group solidarity as it is the very foundation of their economic development success.

These responses reveal a very interesting and significant aspect of the relationship between social and economic empowerment, at least within the context of Proshika's programs and group structure. These A+ groups clearly see themselves as capable of carrying on independently from Proshika in terms of financial/loan support. They appear to be confident in their abilities to succeed in the market place. Yet at the same time, they see their social empowerment as a pre-condition of their economic development success—their strong emphasis of the importance of their group solidarity. They are also prepared to enter into a relationship of "equals" with Proshika —they are willing to pay for technical assistance services provided by Proshika. All these point to the existence of some of the major pre-conditions of sustainable development, anchored on a foundation of strong social empowerment. This would appear to have met the sustainable criterion for an optimal level of social development service.

Case Study 2: Fights for Legal Rights over Protected Forest Land

Bangsi Nagar village under Mirzapur *thana*, Tangail, consists of 22 groups, 12 male and 10 female. At the start of Proshika's activities, there were only four groups (two male, two female) in the village.

After receiving training from Proshika on natural resource management with special reference to Social Forestry, these groups went for forest protection in 1986. Initially, the groups started protecting 80 acres. They started when the area was completely barren, full of *motha* (roots). Now the Sal Forest has again turned green and the trees have reached a height of some 8-10 meters. Later on, in 1982, these groups were involved in roadside plantation and block plantation with Proshika's support and payment in the form of wheat from the World Food Program.

Now 300-350 acres of forests are protected. It is all under the Forest Department, mainly Sal Forest. The protection is against all illegal deforestation within the area—illegal deforestation by

Forest Department officials as well as by outside timber traders and thieves.

By pruning, they collect the thin, small branches as firewood. Using these pruned branches as firewood, they save 20-24 taka a day, the cost of buying the quantity of firewood needed daily from the market. That makes an annual "saving" of 7 to 9 thousand taka.

How do they manage to protect this big area? The group members are appointed to different checkpoints in the forest. They work in shifts for a couple of days. In the day time women group members are responsible for the protection work, while male members cover the night shifts. All protectors are instructed to call out a certain signal as soon as any intruder with suspicious aims enters the forest. At this signal, other members will rush to the spot and interrogate the intruders and if needed chase them away.

In the beginning, they sometimes caught groups of firewood thieves. These were brought to the village *shalish*, controlled by the group members, where they apologized and promised never to try these kinds of tricks anymore. Gradually, these problems more or less vanished. People with bad intentions got to know about the consequences of intruding into the protected forest area and they started avoiding it. Nowadays this problem is minimal.

To begin with, the Forest Department officials were hostile and negative towards the group members' protection efforts. After some years, they realized that the collective work of the group members was a success. The organized villagers proved to be capable of achieving efficient forest protection, something that the Forest Department never had really managed to bring about, although it is an important part of its duty.

The groups have no legal documents on usufructory rights over the protected forest resource. They tried several times to enter into an agreement with the Forest Department regarding their benefit sharing, proposing that 60 per cent of the benefit from the forest products should go to the protecting groups and the remaining 40 per cent to the Forest Department. Although there has been no legal decision as yet, the groups have still retained their spirit of protection.

Among all the groups involved, two groups (one male, one female) are pioneers in the forest protection activities. The activities and the perspectives of these two groups are detailed below.

Ausha Chala Bhumiheen Samity, village Bangshi Nagar, *thana* Mirzapur, was formed 14 years ago. The group started its activities with a total of 41 members. Later on, this large group split into three, leaving 14 members in the original group. The group has so far received four loans for the redemption of mortgaged land, cattle rearing, roadside forestry and nursery project respectively.

For the redemption of mortgaged land, the group received a loan amounting to Tk. 21,500 and obtained a profit of Tk. 10,000 from this project. No significant benefit was accrued to the group from the cattle rearing project. Under the roadside forestry project, the group, with a leasing arrangement, has been taking care of six year old trees along a three kilometer road. From the nursery project, on 20 acres of land, one of the group members earned an incremental income over three years: Tk. 5,000 in the first year, Tk. 13,000 in the second year and Tk. 17,000 in the third year.

As well as gaining provisional access to *khas*[7] forest land, the group has collectively managed to gain possession of an asset of 40 decimal land and has made progress with EIG activities. Besides this, most of the group members have individual savings ranging from Tk. 10,000 to Tk. 20,000.

In addition to economic activities, the group has continuously endeavored to bring about changes in other spheres of their lives. Most of the group members have gone through adult literacy courses and are able to write letters and read books. Some group members have gained access to the committees of the schools in their locality.

Regarding further development of the poor, the group members do not consider economic intervention to be the pivotal part of their activities. As articulated by Iman Ali, the chairman of the group:

"Rin (credit) does not help much in improving the overall condition of the poor. While some income generating activities are prospective, there are some other EIG activities which are vulnerable to natural calamities and unfavorable market conditions. In any project, we have to more or less risk at the implementation level. Contrarily, despite higher social risks, social mobilization can create greater solidarity among the poor which is required for total development. Without social mobilization, it is hardly possible to bring about real changes in our lives".

Ausha Chala Bhumiheen Mohila Samity (Ausha Chala Landless Female Group), one of the oldest groups of the village, was formed 14 years earlier with a total of 13 members. Initially, the

---

[7] *Government-owned resources such as land, forest, ponds etc.*

group contribution was 1 taka per person per week. Later on, the contribution increased to 5 taka. With the gradual progress of EIG activities, the group has acquired permanent assets such as land. For example, they collectively bought 40 decimals of land worth Tk. 50,000.

Before joining the group, most of the members used to work as servants for rich families. At that time, they used to be chased by the rich people while collecting firewood from the forest. After getting into *samity* activities, they have been able to manage their livelihoods by their own savings. In addition, the collective efforts of the group have made it possible to regenerate the big forest land. This has brought about a reversal of the situation. Now, instead of being chased away, the groups have achieved the moral strength to chase away those rich people and others who attempt to cut trees in the big protected forest areas.

Regarding the their own development, the group members predominantly rely on the access and control of their existing resource structure rather than having an absolute dependence on credit. As Duljan, the chairman of the group says,

> "*Credit is always a liability. We are always in tension in terms of loan realization when we do something with the credit money. And whatever we have after repayment is meager. On the contrary, if we gain our access to and control over existing resource base including forest land, whatever we obtain through our collective efforts is our net return. For this, we do not have to think of realization of money*".

Case Study 3: Fights for Legal Control of Khas Resources

In 1986 in the Jangram village, Shibganj *thana*, Bogra there were 200 families altogether: 130 were landless, marginal farmers; 50 were small peasants; 15 were rich peasants and five were landlords, one of whom was considered big, because he had more than 30 acres of irrigated land, and was successfully running a number of business ventures.

The **Jangram Uttor Para Bhumihin Samity**, Jangram North Para Landless Society, was formed in 1981 with 42 members. One of the members has received human development training from Proshika on social mobilization, leadership and organization building strategy. On his initiative another three groups were formed within two years, and now they are preparing to form women's organizations too.

By 1982 there were three landless organized groups in the area and they judged themselves to be strong enough to go for collective mobilization on the issue of increasing their wages. At that time a day labourer worked 11 hours, from 5 a.m. to 6 p.m., for a payment of 10 taka per day plus three meals. Their demand was 20 taka. Finally, after many tough discussions with farmers and landlords and a great deal of tension, they succeeded in getting their claims through.

As labour costs had become more expensive, the landlords put heavy pressure on the organized workers, harassing them during working hours to extort as much labour as possible. As a result, the day labourers met to discuss how to achieve independent standing and reduce the tension and the pressure by the rich farmers and rich peasants, who were buying their labour.

They decided not to sell their labour individually, but instead only to work on a collective contract basis. If any individual was asked to work, he would only accept the whole job, for example to harvest the whole field. The work would then be done by a group of members on a contract basis. This strategy meant two possible gains. First, they could bargain from a stronger position, not as individuals but as a collective effort. No pressure could be put on an individual by the landlords. Secondly, they were working on contract, by settling a flat rate for harvesting a particular field.

In 1983 they organized a fourth *samity*, so that nearly every landless or marginal farmer in the village was organized. A village coordination committee was formed consisting of four representatives from each of the four groups. They started planning to reclaim two *khas* ponds in the village. The ponds were the property of the village but, as often happens, they were controlled by a few of the landlord families.

Just before the monsoon in 1984 the groups took action and went to catch fish in those two ponds. Due to their organizational strength, the landlords did not dare to prevent them, but a village meeting was arranged and the leaders of the coordination committee were threatened, in their absence, by the landlords.

Conspiracies against the group leaders began. The landlords filed a robbery case against two of the coordination committee leaders at the *thana* headquarters. At the same time the landlords arranged a mass-meeting and identified these leaders as "anti-state elements" conducting underground politics in the name of organizing *samities* (groups), calling the group leaders criminals. The house of one of the leaders was searched in their efforts to get hold of him for "special treatment", but he was not at home. The landlords themselves tried their best to terrorize the organized villagers, with the help of the local police.

The village coordination committee arranged another meeting to deal with the conspiracy. The leaders who had been accused of robbery, went to the police station with some Proshika *kormis*[8], claiming that the accusation had resulted from the conflict between the different interests of the village. As a result, the police Inspector asked the Union Parishad Chairman to investigate the case. The chairman organized a meeting between the landlords and the landless in which it became clear that the accusation was false, so it was dismissed. The groups were then granted the lease of the two ponds for three years from the *thana* administration. Both these actions made the groups stronger and their recognition in the area was absolute.

As in all villages, there is a mosque in Jangram, the governing body of which was, naturally, formed by the landlords. Three years earlier there had been a plan to renovate the mosque, for which 10,000 taka was collected from the village people. Work was delayed by the governing body for three years, but then the village coordination committee, supported by poor peasants and small peasants, demanded to see the accounts and make an investigation of the management of the funds. At a village meeting they proved that the governing body was not acting in the public interest. A new governing body was formed consisting of landless members of the organized groups. They recovered 10,000 taka from the old governing body after a couple of meetings, and the renovation work was done within a short time to everybody's satisfaction.

When groups are well organized and comprise almost all the landless and marginal farmers of a locality, they can not only claim their economic rights in the form of increased wages, but can also become a powerful force in social terms, as their strength and unity can make them the leading force for other groups as well. In this case, the example of the poor and small peasants of the village can be taken.

The Jangram village now consists of eight male and four female groups. At present, the ordinary rate for the wage labourers is Tk. 30 plus three meals a day. During peak season, this rate increases to Tk. 70-90 with three meals. In previous years, people used to mi-

---

[8] *Development Worker.*

grate to neighboring districts for agricultural work. Now, this mobility has stopped, partly due to the increasing trend of cropping intensity in the village and partly due to their preoccupation with their own income generating schemes.

Agricultural work is still done on collective basis. Bargaining strength is the same as before, i.e. well developed. The groups have taken lease of two ponds of 2.91 acres and 1.17 acres respectively for 9 years. Out of the two, 8 years have already elapsed for the first one while 7 years for the second one.

In 1992, when the previous *Upazila* system had been reformed, the new Thana Nirbahi Officer (TNO)[9] was reluctant to give an annual extension of the current legal lease, which had already been approved by the Previous Upazila Nirbahi Officer (UNO)[10] for nine years. This resulted in tension among the group members which ended with *gherao* (encircled) of 7,000 organized people, including 2,000 women coming from 13 unions. As a result of this movement, they managed to finalize the leasing arrangement of the said two *khas* ponds dealing with *thana* level bureaucracy without Proshika support.

The groups have taken over and have full control of local public interest groups (*shalish* and mosque committees). Also the unorganized parts of the village fully support the organized group members who have already proved themselves deserving of justice. When Bulu, a leading characters among the organized was threatened with jail, the landless people warned the landlords by saying that if anything happens to Bulu, they (the landlords) would not be left untouched. The groups feel that they have unity and a high level of strength. As they say, "Now it is the landlords who are afraid of us". They have confidence; no *bhoy* (fear), no *lojja* (shame).

When court cases were being brought against the groups, between six and eight members of Jangram Uttor Para Bhumihin Samity left their group. Now the group consists of 17 members. They have a shallow tubewell, and no loan. Presently, the group is planning to receive credit for collective dairy production. The group has mobilized so far savings amounting to Tk. 85,000.

Nowadays, most of the members are self-employed due to their increased economic empowerment whereas before they were fully

---

[9] Chief of Thana Administration.

[10] Chief of Upazila Administration.

dependent on selling their labour. This means a better economic standard which implies less pressure to maintain the traditional system of dowry. Dowry is therefore not considered as a big problem; it is not connected with social pressure. They have achieved 20 per cent dowry-less marriages in their locality. The group feels that dowry-less marriage is fully possible, but it will take time to get rid of this problem completely.

In Jangram, a very high level of social consciousness as well as social mobilization capacity has been developed. The overall empowerment level achieved, both social and economic, is quite impressive. The core values around social empowerment are emphasized. Cooperation and unity are stressed. Much positive economic empowerment has been secured as well. Since organizational efforts started a lot of high quality human development has been achieved. This case study is a good example of the mutual reinforcement between social and economic empowerment.

Case Study 4: Beyond the Dream Position

Saleha is a landless and poor village woman. She has been part of a group for ten years. Her group consists of 18 women, all of them landless. She is married to Mustafa and they have two daughters, Rina 14 years old and Khushi 11 years old.

The group started in a tentative manner. It was an experience they were unfamiliar with. They were shy. But once a week they met in Saleha's courtyard. They started regular savings, in the beginning one taka per person per week.

As the years went by, their inexperience and shyness faded. The weekly meetings were filled with discussions about the problems they shared and the world they lived in. Their insight deepened and realization started to get formulated. Saleha looked forward to every meeting. She enjoyed them and found them interesting. The group was growing as human beings. The regular savings continued; soon it became two taka a week and later five. The first minor investments had already been recovered.

Discussions in the group continued. Part of their savings, together with small credits, were invested in income generating activities. Saleha bought some *maunds* of threshed paddy at the time of *aman*[11]

---

[11] Local Rice Variety.

harvest, husked it and stored it for some time and was able to sell it at a small profit.

Then she turned to poultry-rearing. She started with a handful of egg-layers, selling a few eggs now and then and breeding a couple of batches of chicken. The activity grew until eventually she had about fifty layers. But half of them died, because vaccine was unobtainable. After that experience she kept to poultry on a small scale.

Later, she shifted into goat-rearing for a couple of years, because she thought that by planning in a clever way and providing well-fed goats for marketing at the eve of Eid-ul-Azha, the annual Muslim sacrifice festival, Korbani, she could earn quite a nice sum.

After six years she took a bigger credit to start a cow-fattening project, rearing bull calves into beef cattle. She managed her installments conscientiously and her income became vital to lift her family out of the worst poverty. For two years she has been the owner of two milch cows.

She sells some milk every day and provides her two daughters each with a mug of milk for breakfast. Every year she sells two young stock which means important money for the household. With security provided by Saleha's daily milk-selling income and careful planning of the income, their situation became tolerable for the then unemployed Mustafa, a day labourer.

Saleha is satisfied with her life. *Ruti* (bread), for breakfast and two rice-meals a day—always three meals a day. This is very different from their life ten years ago. She sends her daughters to school, and is very particular about their dress. She cannot complain about her husband. During the last few years they have been able to procure their own hand tubewell, and a slab latrine together with two neighbors. Their house looks neat.

Saleha's example is like a dream. Like the happy ending of a development fairy tale, she wanders out of the jaws of brutal poverty into a blissful land. It is a simple life but a secure, decent life of dignity. She could easily be "a good example" for any organization working with organizing, empowerment, poverty alleviation and micro credits.

One could stop at this point in the tale of Saleha's life; hoping that everything is now sorted out for Saleha. It looks perfect—just what poverty alleviators dream about.

But very soon Saleha and Mustafa will have to marry off their eldest daughter. Then the major part of ten years' accumulated assets could disappear. Ten years of Saleha's persistent and successful endeavors, her whole life's work, the major part of the dream position, will be used for the dowry, which nowadays is one of the most important factors in the increasing economic burden of the already impoverished.

Development efforts seriously aiming at improving the living conditions of the poor in social as well as economic terms, have to put organized resistance and organized mobilization against dowry high on their agenda.

It has been shown among Proshika organized groups' experiences and under the Rural Employment Sector Program (RESP) in Faridpur, that economic empowerment at group level, among other things, can lead to higher levels of dowry as an improved economic situation reinforces the existing system of dowry. Without the support of a new social consciousness and a higher level of social empowerment to enable the poor to fight the stale, traditional, reactionary, male chauvinistic system, there can be no real progress made concerning gender issues such as the dowry system.

How many assaults, wife beatings and maltreatment, divorces, killings and suicides have happened as a result of dowry or the lack of dowry fulfillment? Hundreds and thousands of young girls employed in the garment sector are busy saving money from their meager salaries. For what? For their own future dowry. Female workers accumulate capital, which is readily transferred into the male lap at the wedding. How does this relate to gender perspectives or true development and empowerment from a female point of view?

Those organizations which step back or only fill the air with empty promises, which seem to be fairly numerous, can never claim that it is real sustainable development or empowerment for both sexes they are working for. The fight against dowry ought to be one of the most important issues for all those seriously involved in development, empowerment and poverty alleviation efforts, at least for a generation ahead.

Overall Comments

The Overwhelming Credit Attraction

The safest way for NGOs to get away from total economic dependency on shrinking aid funds is to increase their share of revolving loan funds. Once these revolving loan funds have been lent to the target group and then repaid, they become an asset fully controlled by the NGOs. A real asset when it comes to future economic security and sustainability from the NGOs' point of view. This then becomes the solution for the NGOs in their struggle for economic sustainability for their organization. This is their answer to the donors, who are aiming for temporary support which leads to economic self-sufficiency and a sustainable future beyond dependency on donor support for the NGOs.

There are probably important reasons behind the huge increase of credit services delivered over the last decade. The possibility of reaching the poor, not least poor women, providing them with economic empowerment through credit being invested into income generating activities, without touching or challenging the prevailing social and economical order, is another important reason behind the credit avalanche.

The Grameen Bank success has had a very significant impact. Many donors as well as recipients of aid are in favour of the "soft" approach of credit as a way out of poverty, if such it is. Providing credit to the poor does not provoke any conflict, it is a method which is easily accepted even in conservative quarters nowadays. Especially since the poor recipients of loans prove to be far more reliable credit consumers than the better-off.

Reaching the poor with credit and thereby facilitating their economic betterment, without challenging or provoking any of the structural reasons behind poverty has a very broad-based support in the deeply economicalized development circles since the 1980s. Consensus on this point prevails from former like-minded quarters to traditionally moderate donors and World Bank credit aid is almost universally supported.

Shame, Instead of Empowerment

Credit can certainly play a very important role in the economic empowerment of the poor. But credit is not the only, nor is it automatically, a savior. All credit stories are not stories with a happy ending. When credit schemes turn out to be unprofitable, when invested group savings as well as credits have turned into costly losses, then the empowerment endeavor turns into its opposite, *lojja*, shame.

A defaulting group is a weak group. A defaulting credit hanging over the heads of organized members is a good way for the group to lose confidence in their potential as an agent of change. *Lojja* is an incredibly strong force, far removed from empowerment efforts. Shame means defeat—defeat on the core values and ideas behind human development. Hope and relative strength turn into vulnerability and lack of confidence. Credit failures are not easily repaired. Numerous groups have defaulted, when they were no longer able to function or dissolved.

Credit: Pro-Individualism and Anti-Collectiveness

It is easier, and more secure, to deal with loans on an individual basis than on a collective one from the management point of view. An increasing share of group credits are actively utilized individually by group members.

In mobilization efforts, the groups' experience is that unity, big numbers and solidarity are their strengths. Other groups are needed for this collective strength. You can not mobilize successfully for *khas* resources without the support of many organized in the locality. Credit, on the other hand, can easily promote a very limited viewpoint in this respect. Concerns beyond the economic value or gain of the group are of little or no interest. Even within groups with a very high level of individualism can be found groups of super-entrepreneurial qualities.

A credit-based approach can also promote a business-minded leadership style among the groups, where the leaders' qualities are leaning towards the successful *babsha wallah*, entrepreneur. While good qualities of leadership in the collective mobilization group tend more towards the human development core values. Individualism promotes control rather than broad-based sharing.

Credit—Leaving Social Issues Behind

Even with very effective credit service delivery to the poor, providing them with evident economic empowerment, a number of very important poverty-related areas are not touched. Credit is not a cure-all solution. Access to local resources, social empowerment, fair treatment by local powerful and government officials, fair wages for labourers and so on, are among others, important issues where mobilization and organizational strength are essential for any worthwhile achievements.

Further consciousness raising among group members in combination with mobilization ability is needed, if positive achievements are to be made on vital social issues like dowry, divorce and abuse against women. Credit alone cannot challenge some of the basic and structural reasons behind poverty. Poverty is by definition far more than a lack of sufficient money. If poverty alleviation is really aimed at, one cannot simply ignore unpleasant words like "exploitation" and "inhuman living conditions". Empowerment of the poor is an absolute pre-requisite, their organizational strength and unity and political consciousness, if changes beyond the marginal ones are ever achievable, and if the removal of poverty "beyond the purse" is also an aim. Can this be aimed at without conflict?

Human Development: Social and Economic Empowerment

To what extent is human development, social empowerment and social justice for the poor possible to achieve, if emphasis is barely put on economic issues and credit service?

No positive changes of importance for the poor are possible to achieve or maintain, if there is not a strong and united popular pressure from below. No law or decision in parliament, no matter how pro-poor those at high level may imply they are, will be of any significant value in practice at village or *thana* level, if there is no strong popular movement forcing it to be applied and maintained. Social justice and social and economic empowerment can not be given from above. It has to be taken by organized efforts by the poor themselves, if it is to be sustainable. This implies conflict with the local powers. Thus, organizational and ideological strength as well as collective actions are crucial.

If a road of a non-conflict is chosen, only limited progress in overall human development aims are possible to achieve. Income generating activities can provide *tin bella bhat* (three meals a day). That is good. Service delivery in combination with higher and more secure earning leads to clean drinking water supply (hand tube wells), better sanitation (latrine slabs), better housing (house loans) and higher frequency of children, not least girls, attending education. These improvements are not in conflict with the interests of the local powerful. But are they enough for the achievement of human development and empowerment? Hardly.

If we aim at targets beyond the level of improvements as above, we enter areas of conflicting interests with the local élites. Access to *khas* resources, fair wages, justice in dealings with local powerful as well as government officials can never be achieved, as long as the poor feel fear. Peaceful co-existence implies oppression and exploitation of the poor by the powerful. Only the poor's organizational and ideological strength, unity, big numbers and mobilization ability can make them a match in this conflict. That is their empowerment.

There have been a lot of critical remarks made on credit, or rather the possible pitfalls and limitations in connection with credit or credit-oriented approach. Credit as a means of income generating activities as such has of course a positive effect. Empowerment efforts beyond *tin bella bhat* (three meals a day), are naturally impossible, if the very basic needs are not met. Well-managed credit providing increased income for the target group is very good. But the empowerment gains reached by credit and income generating activities alone are limited to a higher level of consumption, in broad terms. This has to be matched by a corresponding social and political empowerment and this is where the need for thorough efforts when it comes to social and political consciousness building and collective social mobilization come in. This is the "walking-on-two-legs" strategy.

Insights from Case Studies

The two case studies of Jangram, Shibgonj *thana*, Bogra district and Bangshi Nagar, Mirzapur thana, Tangail district give high class evidence that a high level of socio-political consciousness safeguards against individualistic and traditional business-oriented entreprencurship.

They prove that core values and ideas can be kept alive in their group discussions, which does not in any way conflict with their income generating activities. Group co-operation is high on their agenda; social mobilization is a must. As these components have been absolute pre-requisites for the level of empowerment they have reached according to their own experiences. They have a sound scepticism towards credit as a solution to most problems.

Challenges Ahead

Research Challenges

From an analytical point of view, the central questions are those related to the assessment of the linkages and dynamics between social and economic empowerment. A number of questions can be formulated to express some of the challenges ahead.

   i. What kind of empirical evidence is needed to prove conclusively, one way or the other, the dynamics of the linkages between impacts created by social and economic development services? How can such empirical evidence be obtained? What are some of the appropriate indicators?

  ii. Does social development assistance provide a continuous positive marginal impact in terms of economic development? Is there a threshold condition above which social development services yield no additional return in terms of economic development?

 iii. What other criteria, in addition to the two defined above, could be used to assess the level of optimal social development services?

Policy Challenges Socio-Cultural Factors Undermining Intended Impacts of Credit Programs: The Example of Dowry

- Dowry has increased in percentage as well as actual terms over the last decade
- Dowry is causing severe economical setbacks among the poor
- The size of dowry demanded increases when the economic environment permits
- Dowry and action/mobilization against dowry have been on the agenda for a long time. While Proshika members have a lower incidence of giving dowry in comparison with non-Proshika members, it has, however, increased marginally over the past few years.
- Action against dowry can hardly be very successful or sustainable if it is left to the individual or single *samity* to deal with, as the social pressure surrounding dowry is too strong for that. Successful mobilization against dowry needs a high level of social consciousness building and organizational strength.
- Dowry is behind a big proportion of maltreatment and divorce, often leading to social and economic catastrophe for the women, as well as suicide among wives.
- Effective action and mobilization against dowry on a broad base is an absolute must if there is any seriousness behind the

gender issues effort. It is probably the most pro-women policy endeavor of all.
- Action and mobilization against dowry is also an efficient pro-poor policy in general. Successful action against dowry will be a far more important achievement for the vast majority of the poor than large amounts of micro-credit.

Positive Impacts of Social Mobilization

- Wage increases provide all in a locality, not least the hard core poor, with a higher income. A high level of organizational and mobilization developments has proved able to raise wages and maintain the achieved increase.
- Any kind of local or *khas* resource needs a contract and a strong popular movement to make it accessible to the poor.
- Any kind of pro-poor legislation or government decision needs a popular movement at the local level to be applied in reality. Advocacy at the top level is not enough.
- A high level of social consciousness is to be developed further if vital social issues are ever to be challenged, for example gender-related ones. Human development must go beyond the level of increasing access to money or material means of production.

A General NGO-Bank/Credit Institute

- A general NGO-bank/credit institute, purely involved in credit service delivery would have a positive impact on the non-credit areas of work.
- Today a big proportion of the *kormis*' work-load is spent on credit
- Many professionals have far too little time to spend on their areas of expertise. They are to a high extent dealing with credit services instead of developing and exercising their expertise in relation to the *samity's*.
- There is far too little time for any kind of work outside the credit-related areas. This credit focused approach is hampering the achievement of the Proshika objectives.
- A general NGO-bank/credit institute might be the solution to efforts towards reinforcing collective social mobilization building

on *samity* level again as well as successful implementation of core endeavors such as ecological agriculture and vital social issues.

Structural Change

Advocacy at the national level for pro-poor, environmental friendly issues and so on, are relevant. But these efforts have to have a base of popular support and popular mobilization behind them at the micro level. There are stronger biases towards non-conflict and non-confrontation today than there were ten years ago. But some vital areas for combating poverty can never be reached, if a non-conflict road is followed. Who, if not the NGOs with pro-people and poverty alleviation objects as part of their core objectives, is to argue and mobilize for structural changes such as agricultural and land reforms aimed at a higher level of equity and sustainable land use?

Limitations of Micro Credits

Not all the targeted poor are capable of entrepreneurship and management, even with support and training, to allow them to become successful credit and income generating activities performers.

Directing all efforts towards reaching the poor via credit will exclude considerable numbers of the target group, not least among the hard core poor.

A mono-culture of credit as the only means of poverty alleviation effort is not enough. Other means have to be developed and supported as well.

A big proportion of micro credits are, by definition, very small. They are often only enough to direct the credit consumer towards activities with a high level of self-exploitation, not towards sustainable development endeavours.

More and bigger credits should be aimed at, with a special emphasis on the generation of employment outside the family sphere.

Chapter 6

BRAC's POVERTY ALLEVIATION EFFORTS: A QUARTER CENTURY OF EXPERIENCES AND LEARNING

A Mushtaque R Chowdhury, M Aminul Alam

Abstract

BRAC was founded in 1972 in response to a humanitarian relief oriented need. Over the years it has evolved as a fully-fledged development agency with activities spread all over Bangladesh. The major focus of BRAC is poverty alleviation which is addressed through a holistic approach. The components of BRAC's poverty alleviation programme include social mobilization, organization development, health care, children's education, and micro-credit. The distinguishing features of BRAC programmes include promotion of women in the development process, infusion of technology and scaling up of successful programmes. BRAC's poverty alleviation programme is one of the largest in the country, reaching nearly two million households. The paper presents information on the impact of the poverty alleviation efforts in substantive areas such as health, education, women's lives, material well-being and coping capacity of the participants. A special programme targeted at the 'ultra poor' is also discussed. The paper concludes with a discussion on lessons learned.

Introduction

BRAC is a Bangladeshi non-governmental organization (NGO). It was born in 1972 as a response to a humanitarian need[1]. Over the

---

[1] A brief history of BRAC along with a narration of early experiences is available in Annex 1. More details are found in Chen, 1983; Lovell, 1992; Chowdhury & Cash, 1996.

years BRAC has gone through a series of evolutions. It is now one of the largest NGOs in the country. This paper looks at BRAC's poverty alleviation efforts and examines how it is transforming the lives of the poor in rural Bangladesh. Lessons learned in rural development and poverty alleviation over the last quarter century are also highlighted.

Social Mobilization for Poverty Alleviation: A Holistic Approach

The twin objectives of BRAC are poverty alleviation and empowerment of the poor. It works particularly with the women from poorer families whose lives are dominated by extreme poverty, illiteracy, disease and malnutrition (please see Box 1 for BRAC's Mission Statement).

BOX 1
BRAC Mission Statement

BRAC works with people whose lives are dominated by extreme poverty, illiteracy, disease and malnutrition, especially women and children. Their economic and social empowerment is the primary focus of all BRAC activities. Our success is defined by the positive changes we help people to make in their own lives.

Although the emphasis of BRAC's work is at the personal and village levels, the sustenance of development depends heavily on a pro-poor policy environment. BRAC is committed to playing a role at this level through its research and advocacy work. BRAC works in partnership with like-minded organizations, governmental institutions and donors to achieve its ends.

BRAC believes that development is a complex process requiring a strong dedication to learning and to the sharing of knowledge. Our work is based, therefore, on the services of highly committed, competent and serious professionals.

Source: BRAC (1994).

Poverty is looked at in a holistic sense (Figure 1). It is not only a lack of income or employment but a complex syndrome which is manifested in many different ways. In the words of Amartya Sen (1995), "The point is not the irrelevance of economic variables such as personal incomes, but their severe inadequacy in capturing many of the causal influences on the quality of life and the survival chances of people". Addressing poverty thus means taking a holistic approach. Along with income and employment generation, BRAC works for the development of institutions of the poor, conscientization and awareness building, savings mobilization, children's education, health, gender equity, training and so on. But central to these is the creation of an 'enabling environment' in which the poor can participate in their own development, in which the poor are able to perform to their fullest potentials. All of BRAC's efforts are geared to creating such an environment.

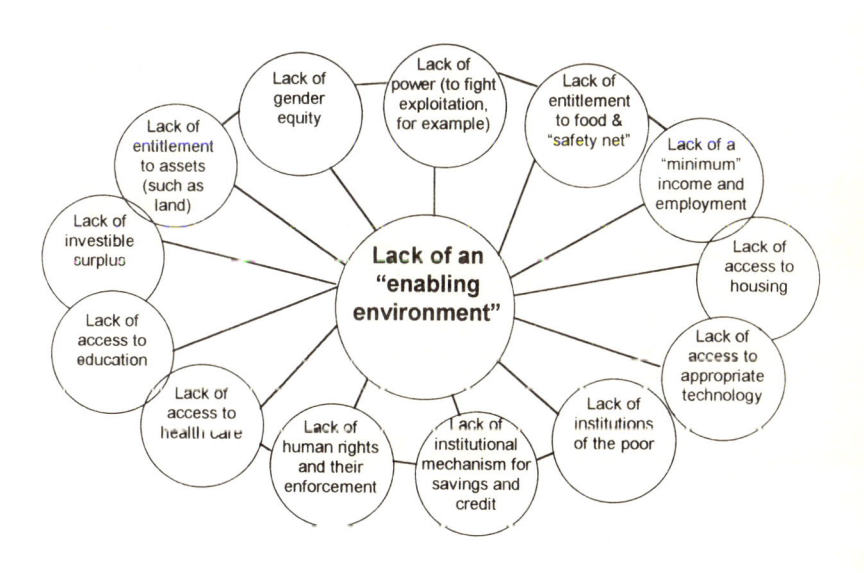

Fig. 1: Causes of Poverty

The Process of Social Mobilization

The process of social mobilization in a village starts with the identification of the poor belonging to BRAC defined target group. Surveyors visit all households in the village to know of their status in terms of land-holding. In recent times, however, BRAC staff have been using selected rapid rural appraisal (RRA) methods including wealth ranking and transect (Chambers, 1992), as additional tools in identifying the target group.

An institution of the poor, called a village organization or VO, is formed as soon as an adequate number of individuals[2] show definite interest. Two activities start simultaneously: a conscientization programme and compulsory savings. Through the conscientization programme, the women are made aware of the society around them. They analyze the reasons for the existing exploitative socio-economic and political system and what they could do to change it in their favour. A formal course on Human Rights and Legal Education (HRLE) is provided to the members[3]. Under savings, the members participate in compulsory saving of at least Taka 2 per week. Savings are considered (by BRAC and group members) as security for old age. Members are allowed to withdraw their savings under certain conditions. Experiments are underway now on devising ways to provide full access to it.

The conscientization process does not stop at a point. This is a continuing educational process which takes place through different fora: weekly and monthly meetings of the VO, training programme of the members at training centres, and the continuous interactions that take place between the VO members and BRAC staff. In each VO, members are trained in different trades. Thus one member may be trained as a village health worker, and another as poultry vaccinator. These cadres cater to the need of VO members and also sell their services to other villagers for a small fee.

Within a month of formation, VO members are allowed to apply for loans from BRAC[4]. Three types of credit are disbursed. The members may request credit for any (a) traditional activity such as rural trading, transport (boat and rickshaw) and rice processing or (b) non-traditional (for women) such as grocery shop, rural restaurant or technology-based activity such as HYV poultry, sericulture or mechanized irrigation; they can also request credit for (c) housing

---

[2] The emphasis on gender has changed over time in BRAC; in the 1990's most VOs formed are composed of women only. A VO has 40 to 50 members, but it can start functioning with 20 members.

[3] The HRLE covers the following: Constitutional/Citizen's Rights, Family Law, Inheritance Law and the Land Law (Abdullah, 1993).

[4] Credit is provided on an individual basis. In the early 1980's BRAC experimented with 'group credit' which did not work well.

loans. While the interest on a housing loan is 10 percent, it is 15 percent for the other types[5]. The repayment rate is 98%. Consciousness on the part of the members, peer group pressure and BRAC staff supervision are the important reasons for this impressive outcome.

BRAC also provides skill training to VO members. An important feature of poverty alleviation activities is that an attempt is made to create a 'backward and forward linkage' for most of the technology-based activities. For example, in the case of poultry programme, BRAC starts with providing training to women on how to rear high yielding varieties (HYV) of chicken. Loans are given for operating a low-cost hatchery to supply day-old chicks to other village women. The women then rear these until they start laying eggs. The eggs are then sold to the hatchery as well as to consumers. One of the major problems of poultry rearing in Bangladesh is the high mortality of the birds. The government livestock department keeps stock of vaccines but these are little used. What BRAC has done in this backward and forward linkage is to train a VO member on how to vaccinate poultry, and link her to the local livestock department of the government which supplies vaccines. The woman receives the vaccine and then inoculates chickens in her own village for a small fee. This way the woman increases her own income and, at the same time, ensures survival of her neighbours' chicken. Similar backward and forward linkages have also been successfully established for other programmes such as sericulture, in which case BRAC also has been able to establish a marketing outlet for the producers through a shop-chain called Aarong[6].

BRAC's Theory of Development: A Learning Approach

BRAC is a 'learning organization', and its learning in the field is constantly used to redefining its programme strategies, which the SAARC Poverty Commission Report has termed as "action-reflection-action" process (SAARC, 1992). Senge (1990) has described the different

---

[5] This rate is higher than commercial rates but appropriate for the following reasons:
   a. BRAC loans are small loans but the processing cost is same whether it is small or big; interest earned from small loan is also small.
   b. BRAC loan is closely supervised and thus the cost is high; BRAC brings banking to the door-step of the poor. The operational cost is thus high.
   c. The rate BRAC charges is required for financial sustainability of the system.

[6] Currently there are 7 branches, six within Bangladesh and one in London, UK.

attributes of a learning organization, and Lovell (1992), Chen (1983), and Korten (1980) have found BRAC as a 'prototype' of a learning organization. Ever since its inception, continuous learning has been the mode of policy planning in BRAC.

The activities of BRAC are steered by several hypotheses, which Lovell (1992) called BRAC's 'theory of development'. The following are some of these hypotheses:

- Every person rich or poor, man or woman, is capable of improving his/her destiny if he/she is given the right opportunity; BRAC believes in the creativity of the poor;
- The participation of women in the development process is essential; People are the subjects not passive objects in development;
- There is no one 'fix-all' approach; the development process must harness the learning culture;
- Conscientization is essential to empowerment;
- Commitment to self reliance should be the goal;
- Small is beautiful but large is necessary; and
- A market perspective and an entrepreneurial spirit are useful.

There are many ways through which the learning culture of BRAC is harnessed and promoted. The monthly project meeting is a major feedback and planning exercise. Staff from a specific programme or field unit, as the case may be, gather together on pre-assigned dates to discuss various matters related to their work. The meetings, often attended by representatives of senior management, discuss successes and failures in their work, set new targets in the light of past experience and debate innovative ways of addressing problems. Such meetings take place at each level: village, area, programme and BRAC. Major strategic decisions are taken in such meetings.

In BRAC, research and evaluation is an important activity. It is becoming increasingly more important as the organization is getting bigger. Such activities are carried out by the Research and Evaluation Division (RED), an independent unit within the organization. With more than a hundred staff, RED carries out specific studies which are both cross-sectional and longitudinal. The main purpose of these studies is to gather information on programmes' performance. In 1995, over 60 studies were carried out by the Division

(BRAC, 1996). At the beginning of each year, RED staff sit with representatives of different programmes to find out about their research and evaluation needs. The results of studies carried out by RED are disseminated through seminars which are held in Dhaka, the headquarters of BRAC, and at field meetings. The reports are circulated to the management. Bangla versions of these reports are also prepared for circulation to the field staff. Traditionally RED studies have concentrated on the 'process' aspects of the programmes, but recently 'impact' studies are being given increasing attention. There are also studies done by outsiders such as academic institutions, the government and donors. BRAC has always been receptive to the findings of such studies and has made strategic decisions in response to these.

The participants of BRAC programmes take an active part in these studies. In one of the recent studies, the programme participants discussed with RED staff the positive and negative aspects of different BRAC programmes and how they weighed or rated them in comparison to other poverty alleviation programmes in their villages. The participants disliked the BRAC policy in respect of participants' access to their own savings with BRAC. As a result, BRAC revised its policy and decided in principle to give group members complete access to their own savings (BRAC, 1995). There is also a Monitoring Department which conducts studies on different programme aspects on ad hoc basis as requested by the programme.

Much investment has been made in developing the capacity of programme participants and BRAC's own staff. A large Training and Resource Centre (TARC) with 14 physical facilities in different parts of the country caters to the training needs of the participants and staff members. Some of the training courses organized in 1994 included 'social awareness education', 'leadership', 'organization and management', for different skill cadres such as poultry vaccinators, village health workers, and primary school teachers. BRAC staff are also sent abroad for training ranging from short-term orientation/seminar/workshop to Masters and PhD level studies.

The Size of BRAC and Financial Sustainability

An important feature of the BRAC Rural Development Programme (RDP), the major poverty alleviation programme of BRAC, is its financial sustainability; the interest revenue from loans covers the operating cost of the programme. Administratively, the programme is managed through branches, and a typical branch would include about 150 VOs or 6,000 members in about 100 villages. At the present rate of expenditure, a branch with an outstanding loan portfolio of Tk. 9 million (US$ = Taka 40) becomes 'self sustaining'. In 1995, about 30 percent of BRAC's Rural Development Programme branches were self sustaining. Table 1 gives some selected facts on the existing size of BRAC's poverty alleviation programmes. By the turn of the century, there will be three million members in 75,000 VOs.

Table 1: Some Basic Facts about BRAC (December 1995)

Full-time staff	16,000
Part-time staff	Over 32,000
Participants in RDP programme	1.5 million (80% women)
Amount of loan disbursed to the poor	Taka 9600 million
Repayment rate	98%
Amount saved by the poor	Taka 800 million
Villages with BRAC programmes	35,000
Total primary schools run by BRAC	35,000
Total students enrolled	1.1 million (70% girls)
Mothers taught on oral rehydration for diarrhoea	13 million
Total budget (annual)	Taka 3500 million
Number of districts with BRAC programmes	58 (out of 64)

BRAC also runs health and education programmes; the main target of these are the VO members. The health programme visited over 13 million households during the 1980s to teach mothers how to prepare salt and sugar solution at home for combating diarrhoea (Chowdhury & Cash, 1996). The present health programme provides essential health services to the villagers through village based health workers who, as already mentioned, are selected from among the VO members. The health programme also conducts pilot programmes to innovate health delivery mechanisms, particularly in respect of women's health.

The education programme, on the other hand, runs 35,000 non-formal primary schools, with 1.1 million students, 70% of whom are girls, for the children of poor families. These are located all over the country and cater to the educational needs of VO members' and other poor children who are otherwise not reached by the formal system.

Salient Features of BRAC Programmes

Promotion of Women in the Development Process

BRAC has been promoting a new culture in the development field with women in the forefront of all activities. Most of the recipients of credit are women; seventy percent of students and eighty percent of the teachers of NFPE schools are female; and health and poultry workers are also all women. Breaking the barrier of a predominantly conservative traditional Muslim society, BRAC has even succeeded in training female workers to use motorbikes in performing their duty. Women are running rural restaurants, vaccinating chicks, treating patients, doing carpentry, or teaching and studying in schools, traditionally all occupations which used to fall in the male domain. This has shaken the base of the traditional power structure of rural Bangladesh, which has felt threatened. Their response has been a backlash against BRAC and other NGO activities in Bangladesh. Over the past years, many of the BRAC programmes have been subjected to physical attack and harassment. Over 50 NFPE schools were burned down, thousands of mulberry trees planted by the poor women were razed to the ground, and many BRAC participants and workers were either harassed or threatened. Although the cover they took was Islam, the forces behind them were those of the traditional power structure—the money lenders, village doctors, and other rural elite—whose realm of influence is diminishing in the wake of this newly emerging force of women and the poor. BRAC has since responded to these attacks by initiating new programmes including increased public relations in the villages where BRAC works.

Infusion of Technology

The greatest potential of micro credit to improve the lives of the poor on a sustainable basis has been offset by a lack of concomitant promotion of technology. Much of such credit has been used for traditional activities, and not enough has been done to include technology along with it. The profit made from traditional activities is not always substantial, not enough to generate an investible surplus. In the case of BRAC, 70 percent of its credit portfolio is occupied by traditional activities. The need for technological infusion is recognized, and BRAC has been making concerted efforts towards this. The Rural Enterprise Project has been working since the mid 1980s to innovate and test possible areas of technology use in BRAC-sponsored income generating activities of the poor. Examples in this respect are: HYV bird, vaccination, hatchery and chick rearing unit in poultry, artificial insemination in livestock, deep tube-well irrigation in crop production, improved varieties of mulberry trees, quality production of cocoons and modern reeling facilities, etc. About 30 percent of the existing credit portfolio is made up of such technology-oriented/intensive activities. BRAC provides training, input supply and marketing support in certain sectors[7]. Such activities will increase the profit margin of the participants, thereby increasing the propensity to save more as well as contribute to boosting the national economy through increased production.

Scaling Up

Small is beautiful but big is necessary, is a quote commonly heard in BRAC. The 'seeds of change' which have been sown need to be multiplied to universalize their benefits and also for the sake of greater impact and sustainability. The following explains why BRAC continued to exist after completion of its initial relief objectives:

> "When BRAC was started in 1972 we thought that it would probably be needed for two to three years, by which time the national government would consolidate and take control of the situation and the people would start benefiting from independence. But as time passed, such a contention appeared to be premature. After 16 years, we feel that we have not yet outlived our utility and need to do more and more" (Abed and Chowdhury, 1989).

The reason why BRAC decided to scale up its programmes may be explained in the same manner. The problems that the people of Bangladesh face are numerous. It is inconceivable that all these problems can be effectively addressed by government alone. The need for others, such as NGOs, to step in is becoming increasingly apparent. In the words of BRAC's Executive Director the following explains why BRAC wanted to expand its activities:

> "To me, with the kind of poverty we have in Bangladesh, with the kind of governmental ability to respond to the needs of our people, we basically have to act. We must act. It is not a question of whether

---

[7] These include HYV poultry, artificial insemination in livestock, sericulture, fisheries and agriculture.

we should or we should not. There is a kind of imperative in the situation that we have in our country. The imperative is: do we serve our people as best as we can out of the kind of poverty, the intense, dire poverty, that we have in our country? Do we remain small and beautiful or do we scale up and try to take the consequences? BRAC has decided to try to take the consequences of becoming large. Hopefully we will try to do the best we can, but there will be problems and we are aware of this and we will try to meet the challenge as it comes along. We have decided as a group that we would like to try to serve as large a number of the poorest people in Bangladesh as we can." (AKF Canada/Novib, 1993).

The notion that NGOs should restrict themselves to small-scale pilot projects and leave it to the government to replicate their more successful experiments is challenged by BRAC. Each project, whether government or NGO, has its own style of implementation, and the pilot projects draw heavily on the organization's own culture and strategies. Replication is best done by the organization concerned, given that it has the necessary capability, resources and the willingness to do it. BRAC believes that its successful experiments can best be replicated by itself.

An example of how an apparently successful small programme failed to keep up the pace when scaled up is provided by the Bangladesh Academy for Rural Development (BARD). The Academy had conducted some very interesting experiments in the Comilla area during the 1960s under the inspiring leadership of Dr. Akhtar Hameed Khan. Important innovations that came out of the Comilla experiment included the 'Two-tier Cooperative System', the 'Thana Irrigation Programme' and the 'Rural Works Programme'. In the early 1970s, the government decided to replicate some of the Comilla experiences all over the country. Soon, a new project, Integrated Rural Development Programme (IRDP), was created and branches opened country-wide. At the time of expansion Khan had already left the Academy and his colleagues were infrequently consulted. Bureaucrats without any direct exposure to or feelings of ownership of the programme were given the responsibility of the fast expansion and the IRDP failed to successfully replicate the Comilla experiment.

Chowdhury and Cash have been looking at the scaling up of BRAC's Oral Rehydration Therapy (ORT) programme of the 1980s which covered the whole country. They identified the following as

the major factors responsible for the successful scaling-up of the BRAC ORT programme (Chowdhury and Cash, 1996):
- learning from the grass-roots ("action-reflection-action" process);
- effective management through innovative management system (such as the incentive salary system for the front-line workers of the programme);
- forward planning;
- planned recruitment of different levels of staff;
- investment in capacity development of staff through local and overseas training;
- imaginative communication drive;
- supportive and effective feedback system and the receptiveness of management towards those; and
- direct support from different important groups such as the government, professional experts and donors.

Helping the Poorest

In spite of all good intentions and efforts, BRAC's success in reaching the poorest ten to fifteen percent of the rural population has been modest. The target group of BRAC are those who own 50 decimals of land or less and sell manual labour for survival. Because of various reasons the people who own marginally more land than the upper limit have also been found to be members of BRAC VOs; they form approximately 15 percent of female VO membership (Mustafa, *et al.*, 1996).

BRAC's standard programme explained above has unfortunately not been successful in reaching the poorest of the poor: the "ultra poor". Because of the very nature of their lives, those who mostly happen to be beggars, women-headed households with very small children or the chronically sick, are unable to participate in many of the 'rituals' of the programme[8] (Evans *et al.*, 1995). They are also less able and willing to use credit profitably and be self employed. Creation of safety nets and wage employment are probably more important in such cases. BRAC has responded to the needs of this group by initiating the Income Generation for the Vulnerable Group Development (IGVGD) programme.

---

[8] The rituals include regular savings and attendance at weekly meetings.

The government, since the 1970's, has been providing a ration of 31 kg of wheat per month to distressed women in rural areas who have no land or income and do not receive support from their husbands because of desertion, divorce, death or disablement. They constitute the bottom 10 percent of the population. Being women in the poorest households they are in most cases excluded from development projects. The programme, called the Vulnerable Group Development (VGD), provides rations for a period of two years, after which a new group is chosen for the dole. BRAC has been working with the government since 1988 in providing training on poultry raising to these women for the period they receive the free wheat. With the help of the government, BRAC also has provided easy credit to them to purchase and rear poultry. The idea is that when the ration is withdrawn after two years, the women can continue to earn an income from the poultry equivalent to the value of the ration. BRAC has also linked these women with the government livestock department to receive vaccines for their poultry. In January 1995, for instance, nearly 325,000 women participated in this programme in 74 *thanas*. Table 2 provides more information on the IGVGD programme.

Table 2: Some Basic Facts about the IGVGD Programme

Particulars	up to 1992	1993	1994	1995	1993-1995
Thana covered	93	93	38	74	74
Trained cardholders	166,740	37,421	53,283	76,214	166,918
Loan disbursed (in Taka)	78,637,250	122,567,500	100,520,120	263,383,724	486,471,344
Loan realized (in Taka)	62,299,924	84,378,325	74,591,267	165,163,984	324,133,576
Loan outstanding (in Taka)	15,807,326	53,996,501	79,925,354	178,145,094	178,145,094
Borrowers	65,075	59,271	56,625	103,211	219,107

Notes : a. In 1992-93 there were about 30 Thanas in RDP and 9 Thanas phased out. In 1994 there were about 20 new Thanas included out of which 18 phased out. At present, (1994-95) there are 74 Thanas.
  b. In 1994 there were about 55,593 trained VGD cardholders in the phased out Thanas. The number is deducted from the current position.

The programme has trained/developed several groups of skilled workers in the poultry sector. The *poultry workers* (PW) are given five days training in poultry management, basic treatment of diseases of the birds, and vaccination. They have been linked with the *Thana* Livestock Office which provides free vaccines. The PWs inoculate birds and provide treatment to sick birds for a small fee. Another group of women has been trained as *chick rearers* who raise day-old HYV (high yielding variety) chicks in chick rearing units for two months. They buy the day-old chicks from hatcheries run by the government and the private sector. The two-month old chicks are then sold to *key rearers*, another group of poor women, who raise ten hens and one HYV cock. BRAC has also provided loans to set up mini hatcheries. To run the mini firm and chick rearing units properly, one *poultry feed-selling centre* has been established in each *thana*. With the spread of hybrid poultry, people are gradually getting used to buying balanced feed from these centres.

A recent study observed that the income earning and asset building component of the VGD programmes was strong but felt that these were weak in supporting women's empowerment (Hashemi, 1996). BRAC is now integrating the IGVGD programme with its mainstream RDP programme after two years (please see later for a discussion on the impact of the mainstream programme).

In recent times, BRAC has started placing emphasis on creating rural employment for the ultra poor. Through a new window, the "poverty graduates" and other entrepreneurs among the poor will be eligible to receive larger loans for initiating rural enterprises. Enterprises which create employment will be given priority. New jobs created through this will hopefully provide new employment opportunities to the poorest who are not able to take advantage of BRAC loans.

Links with the Public Sector and Other NGOs

BRAC works closely with the government. The ways in which the poultry programme of BRAC is linked with the public sector has already been discussed above.

BRAC also works with the government on health, and has been providing assistance in social mobilization for the Expanded Programme on Immunization (EPI) and the family planning programmes.

BRAC also works closely with other NGOs. In education, BRAC assists over 230 small local NGOs in running non-formal primary schools for children. It also works with eight other medium sized

NGOs in developing their management information system (MIS). BRAC believes in pluralism and maintains that all organizations, small or large, should be given the right and freedom to work for the good of the poor according to their own philosophy and strategy.

How Successful have these Efforts been?

The SAARC Poverty Commission Report suggested a framework of analysis which the Report termed 'sociogram', and laid down specific criteria for evaluation of a social mobilization process. According to the Report, the release of the creativity of the poor has to be judged by the "increase in their social consciousness, empowerment and self respect" (SAARC, 1992). To BRAC, as already mentioned, 'poverty alleviation' means not only 'increased income' or 'more employment', but much more. It may now be necessary to discuss how BRAC programmes have addressed and/or affected some of these.

Enabling Environment

The central issue in any poverty alleviation strategy is how an enabling environment is created in which the poor can participate for their uplift and in which the poor are able to release their creativity to the fullest extent. All the efforts that BRAC has taken so far have been geared towards creating this 'enabling environment'. Obviously, society has yet to go a long way to realize this fully; the question is whether the changes are taking place in the desired direction. The following discussion will indicate how BRAC is manipulating such changes.

Institutions: Although there are existing institutions in many villages and at the national levels, none of these voice the needs or opinion of the poor, and they are all controlled by the traditional power elites. BRAC has been trying to create new organizations of the poor, which are designed to cater to the need of its membership. BRAC has formed over 35,000 village organizations (VOs). Whether these organizations are moving towards institutions is a researchable and obviously contentious issue. It depends on what a VO is expected to do. Evidence, however, suggests that the VOs have yet to attain a kind of autonomy through which they will be able to move forward on their own (Mustafa et al., 1996). The challenge for BRAC and other similar agencies is to devise a strategy through which this can be attained.

Education: BRAC has made significant contributions in the field of education among the poor communities. The non-formal primary education (NFPE) programme is providing education to over a million children who would otherwise be left out of schooling. It is not only the children attending the NFPE that benefit, the programme is putting indirect pressure on the public school system to improve its performance thereby having long-term impact on the education scenario in Bangladesh (Ahmed *et al.*, 1993).

Health: As mentioned already, BRAC has taken the message of Oral Rehydration Therapy (ORT) for diarrhoea to all the houses of Bangladesh. Numerous research studies of the programme have recorded excellent dissemination of the message to mothers in all the villages of the country. Recent studies have documented that ORT has occupied a permanent place in the local culture (Chowdhury & Cash, 1996). The recent drop in infant and child mortality in Bangladesh is believed to have been largely due to this. BRAC's involvement in child immunization programmes is another landmark in its partnership with the government. In the areas where BRAC works, the coverage of child immunization is one of the highest in the country.

Gender Equity: From its very inception, BRAC has been sensitive to the question of gender equity. All efforts, be they in credit; children's education; or health; are designed to benefit women thereby reducing the gender gap in the society. Women who have been empowered now exercise greater influence and control than before within and outside their own households. Within the organization of BRAC, gender neutrality is strictly adhered to (Chen, 1983). At present about 25 percent of the staff are female, a figure which BRAC wishes to increase. At the top management level, however, the gap is still wide. For instance, only two of the seven directors are women.

In a recent study on the impact of BRAC programmes on gender relations, members and their husbands were interviewed to understand the change that may or may not have happened in the lives of the women. The following comment of the husband of a VO member may give an idea of the impact that the programme is having on women's control over resources, an important empowerment criterion:

"My wife has the right to buy toiletries and cosmetic jewelry on her own from the income of her loan, even though it is utilized by me.

Last month she bought a cooking pot which cost taka 80. She said that it was essential for her kitchen. I did not argue with her about this necessity. But she would not have done this before receiving loans." (Mustafa et al., 1996).

In another study it was found that women borrowers have full control over their own income in one-third of the cases; in another third both husband and wife have it jointly and in the rest all is controlled by the husband or other male members of the family (Goetz et al., 1994). Hashemi et al (1996) in a recent study found that 60 percent of borrowers in new BRAC villages had 'full' control over their enterprises.

Change in the status of women in the eyes of the greater community has been rather slower. A study on the impact of BRAC programmes did not record much change in the way the community looked at poor women. This was particularly true in the case of the village *Shalish* (court). However, the following comments by a group of VO members reveal the initiation of important changes:

"In the past we used to sit on the ground, with other poor people. And now we are offered to sit on the bench. We don't have active participation in *Shalish*, but we can make plea against the decision if it is taken wrongly."(Mustafa et al., 1996).

Material Well-being: Studies show that women (and men) participating in BRAC sponsored activities have more income (both in terms of amount and source), more assets ownership and are more often employed than non-participants (Chowdhury et al., 1991). A recent study looked at the impact in more details. It indicated a consistent movement along the path to greater wealth and expenditure, and greater impact on less well-off households compared to better-off households (initial endowments). Table 2 shows the performance of BRAC members in terms of two indicators of material improvements: asset accumulation and expenditure. The members did improve their condition and the amount of improvement became more evident as the length of their involvement with BRAC increased. Figure 2 shows the improvement in terms of asset accumulation.

Coping Capacity: The sustainability of increased material income in the short-run is best understood by examining the coping capacities in lean and peak seasons. The available evidence suggests that BRAC members have better coping capacities than non-members and that such capacities increased with the length of membership and amount of credit received (the so called 'critical mass' of BRAC membership) (Mustafa et al., 1996). Figure 3 shows that among the female member households the proportion reporting 'severe deficit' in food security decreased as the loan size increased. Along the same line of increased coping capacity, Table 3 shows insignificant differences between the lean and peak seasons in selected key material well-being indicators among households which stayed with BRAC for more than two and a half years and received more than Tk. 7,500 in loan. The differences were, however, quite noticeable among their counterparts who received a smaller loan or stayed with BRAC for a shorter period (not shown in Table).

Table 3: Material Improvements of Member Households According to Length of Involvement with BRAC

Indicators	1-11 (n=434)	12-29 (n=262)	30-47 (n=389)	48+ (n=221)	Comparison Group
Average value of gross H/h assets (Tk.)	10,959	14,037	20,282	23,230	7,250
Average % (& value) of assets which are productive (revenue-earning)	32.9 (3,606)	39.1 (5,488)	31.6 (6,409)	31.0 (7,201)	–
Average H/h weekly expenditure (Tk.), including peak & slack seasonal data	419	455	560	528	382

Source: Mustafa et al. (1996).

Table 4: Seasonal Differences in Selected Indicator Values for Female Member Households Staying in BRAC for More than 2.5 years and Receiving Cumulative Loan of More than Tk. 7,500 (n=153)

Key indicators (per capita/per week)	Lean	Peak
Rice consumption (gm.)	3,258	3,062
Food expenditure (Tk.)	65	68
Cash earning (Tk.)	64	59

Source : Mustafa et al (1996).

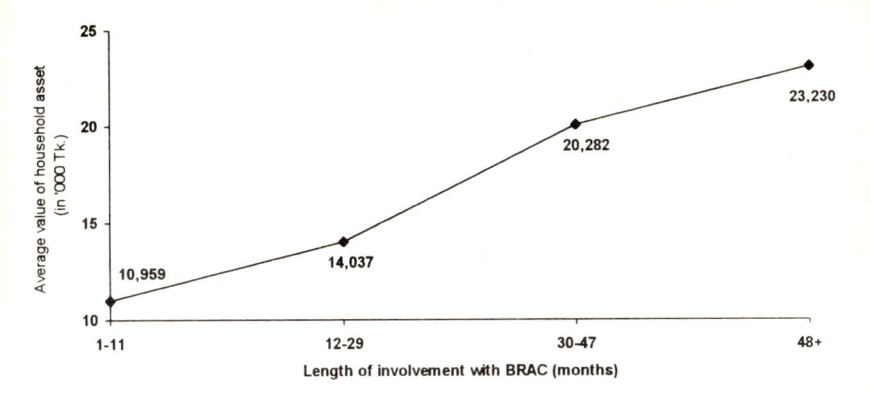

Fig. 2: Increase in the Value of Household Assets Owned by BRAC Members According to Length of Involvement with BRAC

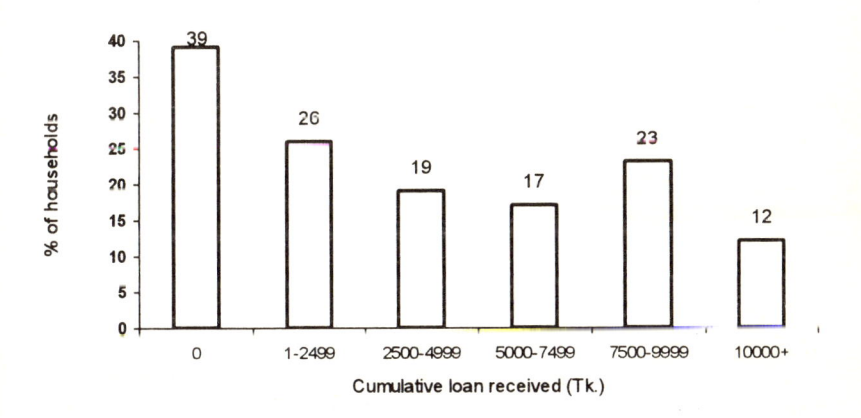

Fig. 3: Proportion of BRAC Female Member Households in 'Severe Deficit' Category According to Cumulative Amount of Loan Received

Environment/Sustainable Development: In terms of sustainable development, BRAC programmes have been a unique success. As an example, the Sericulture programme planted 17 million mulberry trees over the past few years. Apart from adding to a healthy environment, this provided employment to thousands of women in different stages of silk production, garment manufacturing and marketing. In addition, the Social Forestry Programme and the Sanitation Programme are making important contributions towards a healthy and sustainable environment. BRAC has recently created an environment cell to look at environment aspects more systematically.

Discussion

BRAC was born nearly a quarter century ago soon after the Independence of Bangladesh. The task ahead was relief and rehabilitation of the refugees returning from India. After working for a year, it discovered that relief was not a permanent or long-term solution to the multifarious problems faced by the people. BRAC thus started its community development activities. Through this, the whole village was mobilized.

Soon the limitations of that community development approach became obvious. We found that development efforts hardly reached the poor and other disadvantaged sections such as women and fishermen (Chowdhury & Cash, 1996; Lovell, 1992).

The obvious shift from this experience was to what is now known as the 'target group approach'. Through this only the poor are beneficiaries of BRAC programmes. In recent times, more refinements have been made in order to extend benefits to the poorest of the poor.

BRAC is now one of the largest NGOs in this country, with nearly 50,000 full—and part-time staff (Table 1). Let us now move on to the main theme of the paper: the lessons learned.

A most important lesson is that a holistic approach is needed to successfully address poverty. To us, poverty is not only a lack of income or employment (Figure 1). Addressing poverty thus means taking a holistic approach. BRAC promotes institutions of the poor, and works for awareness building, legal and primary education, skills development, savings mobilization, gender equity, health and environment, in addition to income and employment generation through credit.

BRAC's credit programme is large. Last year alone, nearly US$ 100 million was disbursed with a repayment rate of nearly 98%. We have learned that credit is very important and so is credit discipline. Although credit is crucial, it is not sufficient to eradicate poverty. Credit-only interventions may bring impressive results in the short-run but other interventions such as education, health, women's development, etc. are equally needed to sustain improvements. BRAC is also aware that credit discipline must be ensured along with the building of healthy institutions for the poor.

The next lesson is the importance of technology infusion. We feel that the full potential of micro credit to improve the lives of the poor on a sustainable basis has been offset by a lack of concomitant promotion of technology. Thirty percent of BRAC credit now goes to technology-oriented activities. The profit margin from such activities is higher, resulting in increased savings for participants; the ability to give service-charges to BRAC; and participation in increased national production and economic growth. The importance of establishing 'backward and forward' linkage in sustainable poverty alleviation programmes is another lesson for BRAC. BRAC has been trying to establish this linkage for most of the technology-based activities such as poultry and sericulture.

A contentious issue not always recognized in micro-credit programmes is their success in reaching the poorest—the 'ultra poor'. BRAC's standard programme unfortunately has not been able to reach all such households in the villages it works in (Evans et al., 1995). Because of the very nature of their lives, the ultra poor are unable to participate in many of the programme activities regularly. They are also less able to use credit profitably because of their particular disadvantage, namely extreme poverty. For them, the creation of safety nets and wage employment are logical interventions. BRAC has been experimenting with different ways to bring them under its programming. One of the more successful ones is the IGVGD programme, described in the text above. BRAC has also started what it calls enterprise loans for the "poverty graduates" and other not-too poor households which will generate employment for the poorest.

The next lesson is related to women's participation in the development process. BRAC has been promoting a new culture in the development field with women in the forefront of all activities. Breaking the barrier of a conservative society, women are running restaurants, treating patients, going to schools or doing carpentry, all of which traditionally fell in the male domain. Promotion of women in this way has had its obvious backlash from the vested interest groups. But BRAC has moved on with its own agenda undeterred by any threats, whilst respecting the local culture in which the programmes operate.

The next lesson is related to pluralism in development. In Bangladesh, many poverty alleviation interventions are undertaken by NGOs. Because of their special characteristics, they are better able to reach and mobilize the poor and provide services. The government also has a very special role, and collaboration of government and NGO sectors can augment the poverty alleviation process. BRAC has quite successfully worked with the government in several areas. Similarly, coordination and, collaboration and where possible, pooling of resources of NGOs may bring further success. Bangladesh needs many more NGOs. Healthy but coordinated competition among them will bring innovations and ensure better services to the poor.

BRAC believes in scaling up its more successful experiments. It recognizes that 'small is beautiful' but believes that 'large is necessary' (Chowdhury & Cash, 1996). The notion that NGOs should restrict themselves to small-scale pilot projects and leave it to the government to replicate their more successful experiments is challenged by BRAC. It intends to serve as large a number of poor people in Bangladesh as it can.

The last lesson is that poverty alleviation programmes as run by BRAC can have a solid impact on the poor. There is convincing evidence, some of which is provided in this paper, that suggests that involvement with BRAC changed the lives of the participants in many different ways. They are better-off in terms of material well-being; they have more income, more employment and better coping capacities. Their education levels have improved, their health and nutritional statuses are better and the women have an improved status within their own families and in the society they live in. Although BRAC has not yet been completely successful in reaching the 'ultra poor', some activities are reaching out to and benefiting the poorest of the poor. BRAC is also trying to integrate these special programmes (such as IGVGD) with its mainstream programme to bring the poorest to the centre of all development activities.

Annex 1

GENESIS AND EVOLUTION OF BRAC PROGRAMMES

On November 12, 1970 Bangladesh was struck by one of history's worst natural disasters. A severe cyclonic storm accompanied by tidal surges swept the coastal belt of what was at that time East Pakistan, now Bangladesh. Half a million people lost their lives. This was a turning point for the land as well as many a people living in this deltaic plain. On March 25, 1971, the Pakistan Army cracked down on the citizens of East Pakistan, voiding an earlier election results. As a result ten million people sought refuge in neighbouring India. On December 16, 1971 the Pakistan Army surrendered and Bangladesh became independent. Soon the refugees started returning to their homes only to find a barren piece; their homes destroyed or burned down and belongings looted.

In some inaccessible areas such as Sulla in Sunamganj district (formerly Sylhet) with a population of over 200,000, the situation was even worse. The new government in Dhaka was faced with numerous problems and it was too difficult for them to provide the much needed relief to the returning refugees of Sulla. A group of Bangladeshis, some of whom came from that part of the country and who worked together in providing relief to the survivors of the 1970 cyclone, got together again to provide relief to the people of Sulla under a new organization called *Bangladesh Rehabilitation Assistance Committee* or BRAC. The group gathered around them a core of young university graduates and started carrying out relief and rehabilitation work in 200 villages. Working closely with the people, over 14,000 houses were rebuilt, tools such as looms, wheels, hammers, saws, and chisels were given to crafts people to help them resume their respective trades.

The first year was a very valuable experience as BRAC realized that relief measures were fine as short-term goals, but did not take care of the longer-term needs of the people. BRAC also realized that "the poor are rich" in their ability to face insurmountable odds. The change in the name to *Bangladesh Rural Advancement Committee*, with the same initials, was in response to that realization.

This was a major shift in the organization's philosophy, a shift in its emphasis from addressing an *acute crisis* to a *persistent crisis*[9]. The experiences of the relief phase in the first year gave BRAC staff a solid base to pursue development, and a number of programmes were fielded. Most of these early programmes were adapted from other experiments or efforts in or outside of Bangladesh. Notable were the cooperatives from the Comilla model of development and the barefoot doctors of China. The experiments involved the whole village community with some special emphasis on women and included functional education for adults, village group formation, agriculture development, and health and family planning. More details of the early programmes are available elsewhere (Lovell, 1992; Chowdhury & Cash, in Press).

The community development phase (1973-77) addressed the problems of all villagers, rich and poor, men and women, with the village as a single entity. Critical analyses by programme staff as well as findings from research studies began to unfurl the innate groupings within a village community based on faction, kinship, and affluence. Bringing conflicting interest groups under one umbrella of a village cooperative to improve the conditions of those disadvantaged proved to be impossible. This led BRAC to adopt a 'target group' approach. Under this new approach, people belonging to the disadvantaged sections of the community such as the landless, fishermen and the women became the target recipients of BRAC interventions. Since then all BRAC programmes are targeted to the poorest households and to their woman members. The exact definition of 'target group' has undergone several refinements. Households owning less than half an acre of land and selling manual labour for survival are now considered BRAC's target group. More than half of Bangladesh's 112 million population would fall into this category. It is interesting to note that most of the poverty alleviation programmes run by nongovernmental organizations (NGOs) in Bangladesh are similarly targeted.

---

[9] BRAC, however, did not say goodbye to relief altogether. Whenever a new crisis situation arose BRAC responded to it with all its energy. This happened during the famine of 1974, the various floods including those of 1987 and 1988, and the cyclones of 1985 and 1991.

Chapter 7

THE ASA "SELF-RELIANT DEVELOPMENT MODEL"

A K Aminur Rashid

Introduction

ASA's steadfast belief in self-reliance stems from the idea that lack of control over resources is lack of control over destiny. This is as true for the organization dedicated to empowering the poor as it is for the poor themselves. This philosophy has propelled ASA to a position as one of the world's largest financially sustainable NGOs in the field of micro-finance with increasing influence in development programs world-wide. ASA is unique because it merges financial sustainability *and* empowerment of the poor by acquiring the ability to cover a large number of disadvantaged women with quality financial *and* social development services. ASA is sustainable at the local program level as well as the organizational level, meaning complete termination of foreign assistance would not affect current services provided to the poor.

Post-independence Bangladesh found relief missions and development projects meeting generous flows of foreign funds along with a top-down bureaucratic approach to development. In this context, the development of a grassroots organization like ASA based on a self-reliant model is revolutionary. Having nearly two decades of experience in addressing the needs and desires of the disadvantaged, ASA offers many NGOs an alternative to financial dependency which leads to inefficiency, lack of ingenuity and autonomy, and ultimately the retraction of services when funds dry up.

ASA began in 1978 geared for empowering the oppressed through "peoples' organizations", mobilized for social action against exploitation and through legal aid to fight social injustice. In 1984, ASA saw a shift in focus to the basic social unit—the family, recognizing the critical role women must play in development. In this phase, development education pertaining to the everyday challenges of an impoverished household became a continuous process of empowerment through the tangible benefits of improved health, nutrition and sanitation.

ASA has a unique capacity to adapt and build on previous successes. In the late 80s, ASA began to incorporate management skills for income generating projects and stressed the importance of savings in development education efforts. At this point, credit delivery came as a quite natural extension of its success with its development education to rural poor women. ASA was able to tailor a credit delivery model to the needs and desires of several thousand women anxious to escape the poverty cycle. This approach is very characteristic of ASA and is the key to understanding how ASA has transformed itself into a managerial dynamo whose business is serving the poor while at the same time achieving financial self-reliance.

The ASA model is based on decentralized decision making, standardized operations, discipline and efficient use of funds making it competitive even by private sector standards. Simple replication of this model allows ASA to expand outreach to the rural poor with amazing rapidity; presently it involves 550,000 rural poor women, has plans to reach 800,000 members by the year 2000 nation-wide, and has a world-wide influence on credit delivery programs.

ASA's Internal Organization

The organization of ASA is characterized by a relatively small central office in Dhaka, with a Chief Executive, about 50 staff members and roughly 500 branches or unit offices serving 500,000 rural poor women throughout Bangladesh. The unit office is the basic element of the organization at the field level for the formation of groups, development education, collection of savings, loan disbursement and repayment. On average it comprises 60-72 groups of about 20 female members. Each unit has six employees: one Unit Manager (UM), four Community Organizers (CO) and one peon. The COs are responsible for the work with the groups and are under the supervision of the UM. For 10-12 units there is one Regional Manager who serves as supervisor and communication link from the field to headquarters. In the central office one person is assigned to serve and supervise the activities of 3-4 RMs, or 30-40 units.

In some cases, in the remote areas where a full size 'Unit Office' is not possible to establish, ASA allows structural flexibility by adopting a 'Sub-Unit' (with two COs) or a 'Satellite-Center' (with only one CO) to reach the target population.

Financial Flows, Operational and Financial Self Sufficiency

By starting one unit office in an area it is possible within 3-5 months to achieve full capacity—1440 members in 72 groups. Loan disbursements start after 3 months of group formation and the process is completed with total loan disbursements of Tk. 48,60,000 within one month by giving loans to 5 members per week. Through this process loan disbursements are completed within 3 months after lending is started and within 6 months after the start of the unit office. Loan repayments with a 12.5% service charge added are made through 45 equal installments over one full year, including holidays and grace periods. The initial loan size is Tk. 3,000 (maximum) with a Tk. 1,000 increase (optional) per loan cycle. Initial and successive loan sizes can be less than the maximum limit depending upon the borrower's capability. Loan approval and disbursement authority along with all other aspects of the units financial management is delegated to the unit level.

From weekly repayment installments on the initial Tk 3,000 loan, income begins in the 4th month and by the 6th month at Tk. 36,000 exceeds monthly unit costs at Tk. 22,545. By the end of the 9th month the total income of Tk. 204,000 exceeds total expenses (including central office overhead and financial costs of one unit) during the same period (Tk. 203,000).

Through revolving funds the units are able to manage operations after only minimal initial capital investments of Tk. 6.5 million from the central office. The amount recovered through weekly installments from the first unit is taken to the second unit as capital and distributed as loans. Capital revolves in this fashion, meeting the investment needs of four unit offices totaling Tk. 17.28 million.

Calculation of Unit Self-Reliance for Initial Year

Initial credit disburse Tk 3,000 Service charge 12.50 %

Month	Mem-bers	Credit disburse	Income (Service charge)	Cumulative income	Expenditure incl. interest on capital	Cumulative expenditure	Break even month
(1)	(2)	{3(2 × Tk 3000)}	(4)	{5(4+5)}	(6)	{7(6+7)}	{8(5-7)}
1	0	0	0	0	22,545	22,545	−22,545
2	0	0	0	0	22,545	45,090	−45,090
3	0	0	0	0	22,545	67,635	−67,635
4	480	1,440,000	4,000	4,000	22,545	90,182	−86,182
5	480	1,440,000	20,000	24,000	22,545	112,727	−88,727
6	480	1,440,000	36,000	60,000	22,545	135,272	−75,272
7	0	0	48,000	108,000	22,545	157,817	−49,817
8	0	0	48,000	156,000	22,545	180,362	−24,362
9	0	0	48,000	204,000	22,545	202,907	1,093
10	0	0	48,000	252,000	22,545	225,454	26,546
11	0	0	48,000	300,000	22,545	247,999	52,001
12	0	0	48,000	348,000	22,545	270,544	77,457
Total	1,440	4,320,000	348,000	348,000	270,544	270,544	77,457

Beneficiary Self-Reliance

Ultimately, ASA's 'Self-reliant Development Model' is only successful if its beneficiaries become self-reliant themselves. To lead the beneficiary to a state of "self-reliance" themselves, involvement for at least ten years in an income generation program is needed. Continuous investment and re-investment for a reasonable time enables a person to accumulate a reasonable amount of capital for running a small business independently; fund accumulation results in people's financial independence. Observing this trend of fund accumulation by the beneficiaries, ASA has extended the duration of support and services to ensure the beneficiaries' economic stability.

Common income generating activities include rice-related activities, raising livestock and making handicrafts. It has been observed that the borrowers invest their credit amount in small projects with quick regular return which they themselves select and operate. Below are examples of typical paddy husking and puffed rice projects, most common among members.

Paddy Husking

It is found that three days are needed to complete one operating cycle with 5 mounds of paddy, or 10 mounds per week. Investors usually get Tk. 25 per mound after meeting all operational and marketing costs. From gross profit of Tk. 250 per week, after members make the Tk. 75 repayment installments, Tk. 10 weekly savings deposits and cover miscellaneous expenses, families typically accumulate Tk. 30 in cash from operations, or 1% of the invested amount of Tk. 3000 in the first year. A borrower generally conducts business activities 40 weeks per year.

Paddy Husking Project Example

Year	Previous Year's Capital	Current Loan Amount	Total Capital	Actual Invest. Amount	Total (1% × 40 wks.)	Loan Repayment	Weekly Savings Deposits	Total Cash Accum.
1	0	3000	3000	3000	4200	3375	490	335
2	3000	4000	7000	4500	6300	4500	490	1310
3	4000	5000	9000	6000	8400	5625	490	2285
4	5000	6000	11000	6000	8400	6750	490	1160
5	6000	7000	13000	6000	8400	7875	490	35

Due to the labor-intensive nature of a paddy husking business, a borrower usually finds other investments to supplement the activity when about a Tk. 6000 investment in paddy husking is achieved (thus the decline in net profits as increasingly higher loan amounts require higher repayment amounts).

Puffed Rice

Another project in great demand, especially in those areas with access to urban markets through middlemen, is puffed rice. The borrower usually goes to the village market twice a week, 40 weeks per year to conduct her business. Cash is accumulated at roughly 4% per week on the invested amount after paying all relevant costs along with miscellaneous family expenditure.

Through this process, the initial and subsequent investments turn over considerable profits leading to a relatively stable long-term economic position for the investor; family expenditure and overall living conditions improve remarkably. Savings over five years, along with the accumulated profits as a supplement to the main family income go a long way towards achieving family self-reliance.

Puffed Rice Project Example

Year	Previous Year's Capital	Current Loan Amount	Total Capital	Actual Invest. Amount	Total (4% × 40 wks.)	Loan Repayment	Weekly Savings Deposits	Total Cash Accum.
1	0	3000	3000	3000	4800	3375	490	935
2	3000	4000	7000	5000	8000	4500	490	3010
3	4000	5000	9000	6000	9600	5625	490	3485
4	5000	6000	11000	7000	11200	6750	490	3960
5	6000	7000	13000	8000	12800	7875	490	4435

Sustainability Concepts

Institutional sustainability is the willingness and capacity of the organization to take full responsibility for continuing positive changes. To strengthen its institutional sustainability, ASA has taken strong measures to organize groups with effective criteria for group formation. In this respect, husbands and other family members of group members were consulted and their assistance requested to form local groups as well as to build up the basic foundation of the local unit offices. To raise the capacity and level of awareness of the group members, leadership development training along with life centered development education was imparted. Moreover, group members were motivated to maintain group discipline and participate in the group whole-heartedly so that they could maintain regular contribution through the service charges for administrative expenses and regular installments of the principal amount for smooth functioning of the credit program. To strengthen the activities of the local unit offices, policy decisions were taken to maintain regular contact with local NGOs and government to avoid duplication and unhealthy competition or hazards created by the anti-development elements of the society.

Behavioral sustainability is demonstrated knowledge or skill developed as a result of organizational involvement. Keeping in mind the long-term impact of ASA's development initiative, the organization concentrated on bringing changes in the behavioral pattern of the beneficiaries to attain behavioral sustainability. To scale up the confidence level of rural disadvantaged women ASA concentrated on several areas. One area was motivation for self-development through leadership development training at the grass roots level. Another was the creation of a suitable environment for the realization of self-development through group formation, education sessions and other social activities. Strengthening of family relationships through gender awareness training, particularly men-women relationships and the implication of such relationships for attaining a sustainable status was also stressed. Finally, ways of dealing strongly with social issues such as dowry, polygamy, early marriage, divorce, etc. were concentrated on.

Life-centered messages related to health, nutrition and environment were also introduced through regular weekly education sessions for improving the women's standard of living. They are also motivated to take advantage of family planning, health services and nutritional facilities out of their income generating projects. Increased income of the poor is spent on consumption, schooling, housing, etc., and their quality of life has been improving gradually. This reflects the behavioral changes of the poor towards life.

The main concentration of these motivational techniques was education at the individual level as individuals can then deliver their messages at the family and community level. This technique was followed to develop regular behavioral changes to achieve a state of behavioral sustainability. In this respect, a study on "people's attitudes towards people's power and resources" has been conducted to assess the attitude, level of capacity, knowledge and confidence of the rural women and judge whether they would be able to bear the responsibilities without direct support from ASA. A series of workshops on the formation of peoples' committees and network building was also arranged for assessing the strength and weaknesses of the beneficiaries for providing necessary input to attain behavioral sustainability.

Policy sustainability is the acceptance and replication of policy by other organizations, such as government, donors and other national and international institutions. To achieve a higher state of sustainability, ASA maintained a flexible strategy in policy formulation. Through this strategy ASA successfully attracted the attention of local people as well as the new financial institutions in strengthening its institutional resource base for attaining financial stability. Contributions from government financial institutions has also increased over time. ASA has developed a base for new partnership with the Netherlands government, the Agrani bank and the World Bank. Recently, massive expansion has taken place in remote areas. ASA has also provided technical assistance for micro-finance to Tajikistan, Afghanistan, Pakistan, Jordan and Ethiopia.

Chapter 8

CREDIT FOR POVERTY ALLEVIATION IN BANGLADESH: PERFORMANCE OF PUBLIC SECTOR BANKS

Mosharraf Hossain Khan

Introduction and Background

One of the main thrusts of the Bangladesh Government in development planning since the country's independence has been to attain self sufficiency in food production to save the millions in hard-earned foreign currency spent on importing food grain. The role of the financial institutions, especially the rural branches of the Nationalized Commercial Banks (NCBs), the Bangladesh Krishi Bank (BKB) and the Rajshahi Krishi Unnayan Bank, have been very instrumental in achieving increased production and in developing the rural economy as a whole. But their contribution in dispensing credit for poverty alleviation has always been negligible in comparison to that of the NGOs in terms of coverage and outreach.

The Private Voluntary Organisations (PVDOs) or the Non-Governmental Organisations (NGOs) and the Grameen Bank have made commendable achievements in poverty alleviation through creating income-generating opportunities for the rural poor with special focus on women, although their contribution in increasing farm production is negligible compared with that of the banks.

However, their performances in creating employment opportunities for the rural poor and alleviation of poverty are considered far better than those of the banks on the following counts:

- High rate of recovery of credit ensured through close supervision of end-use. The recovery rate of loans disbursed by these agencies ranges from 95-99%.
- Credit preceded and followed by strict group discipline (regular weekly meetings, savings mobilization, repayment of weekly instalments etc.) and motivation/training on various aspects of social discrimination, human development and functional literacy.
- Credit to cohesive groups preferred in contrast to the banks' individual approach. Peer pressure, the main theme of the group approach, acts as a driving force in ensuring timely repayment of credit.
- Credit preceded by skill development training, where necessary, and followed by marketing support to the entrepreneurs—the banks make no such provisions.
- Credit linked to a goal-oriented savings program with a view to generating beneficiaries own investible fund, thereby gradually lessening their dependence on credit.
- Credit followed by Health and Sanitation program to improve the overall living conditions of the beneficiaries.

The Role of the Banks in Dispensing Credit for Poverty Alleviation

Although the banks' rural lending programs have concentrated mainly on financing crop production and long-term lending for irrigation equipment, they have also a number of programs targeted towards uplifting the economic condition of the rural poor. This paper aims to throw some light on the various poverty alleviation credit programs run by different banks. The paper is divided into three sections: the common programs of all banks; the programs implemented individually by the banks; and issues and recommendations.

Common Programs of All Banks

The poverty alleviation credit programs run by the **Swanirvar Bangladesh** is the most widely-known poverty alleviation credit program in which all the banks are involved. It should be mentioned that the Swanirvar Credit program is one of the oldest credit programs run in Bangladesh and it follows the same modality in respect of group formation and savings mobilization as the Grameen Bank. It is also noteworthy that the Swanirvar Credit program was preceded by the "Dheki Loan Program" (Paddy Husking Credit Program) launched by all the banks at the instigation of the Bangladesh Bank in 1978. The banks also handled the "Grameen Bank Project" before it became an independent bank. So, the involvement of the banks in the Dheki Loan Program and the Grameen Bank Project may be considered an "ice breaking" experience for the banks in giving collateral-free credit to the landless people. The Swanirvar Program, although running for more than a decade since 1978, has not reached the commendable level of the Grameen bank and other NGOs in respect of loan recovery, for various reasons. One reason might be the institutional weakness of the program. It should, of course, be noted in this connection that unlike the Grameen Bank and other NGO interventions, the Swanirvar Program has very little support from donor agencies. Efforts have now been launched towards institutional strengthening of the program through the active collaboration of the banks involved—the NCBs and the BKB/RAKUB. The program covers an area of 1024 unions and more than 10,000 villages in 135 *thana* of 39 districts throughout the country with about 0.65 million borrower members. A sum of taka 137.10 *crore* has been disbursed through 700 bank branches. The banks' figures for loans disbursed and recovered are shown in Annex 1.

The **Marginal and Small Farms System Crop Intensification Project (MSFSCIP)**, implemented in Kurigram district in collaboration with the IFAD and GTZ, is another poverty alleviation credit program which involved all the banks. This project differed from scores of other poverty alleviation credit program in the following ways:

- the project was aimed at testing a model of linkage between the banks and Self Help Groups (SHGs) with the active collaboration of other agencies (NGO and DAE)
- the target group comprised two groups—marginal/landless farmers and small farmers
- the project contracted a local NGO (RDRS) and the DAE to undertake the social mobilization part of the project on

behalf of the landless group and the small farmers group respectively
- the donors provided training and other logistic support to the involved agencies.

The Project ended on June 30, 1996 with a recovery rate of about 97% and the banks have accepted the outcome as a viable model replicable in other areas. The Rajshahi Krishi Unnayan Bank—one of the banks involved—has already started replicating this model in Rajshahi area under its own initiative. The banks' information on groups formed and loans disbursed under this project is shown in Annex 2.

Programs Implemented Individually by the Banks

Sonali Bank

Sonali Bank, the largest of the Commercial Banks in this country with 1308 branches (seven overseas branches) moved into rural lending in 1973-74. Present outstanding rural credit stands at Taka 18.35 billion—the major portion of which has gone to boost agricultural production and to help small potential entrepreneurs to undertake various income generating activities. The Sonali Bank's rural credit is being channeled through 1020 rural branches in 1140 allocated unions out of 4451 unions of the country. The total number of rural credit borrowers with outstanding loans stands at 1.04 million. This excludes a similar number. of about 1.27 million beneficiaries (0.64 million male and 0.63 million female) served by the Bank indirectly through the BRDB-TCCA system. The bank's exclusive poverty alleviation credit programs include the following:

The **Rural Poor Cooperative Project** is being implemented in 82 *thanas* of greater Rajshahi, Pabna, Kushtia and Jessore districts with financial assistance from Asian Development Bank. The project aims to create employment opportunities for rural landless poor men and women by providing them with credit preceded by skill training. The project was started in 1994 and about 0.17 million rural poor men and women have so far been given credit of Taka 606 million. The rate of recovery is about 100%.

The **BARD-Sonali Bank Functional Research Project** credit program was launched in 1994 by the Bangladesh Academy for

Rural Development (BARD) and the Sonali Bank. The project aims to identify a viable model for increasing the income of beneficiaries through bank credit as well as ensuring recovery of loans; field level coordination and supervision is undertaken jointly by the bank and the designated officials of BARD. The Project area includes selected *thana* of Comilla, Narayanganj and Sylhet districts. A sum of Taka 2.6 million has so far been disbursed to 900 target beneficiaries and the rate of recovery is 100%.

A similar collaborative project, the **RDA-Sonali Bank Functional Research Project** was also been launched with the Rural Development Academy (RDA) in 1994 covering the selected *thana* of Bogra, Gaibanda, Jhenaidah and Kushtia district. A sum of Taka 1.6 million has so far been disbursed to 600 beneficiaries and the rate of recovery is 100%.

The **Bittahin Rin Prakalpa (Landless Credit Program)** was launched in 1984 in collaboration with the Bureau of Manpower Employment and Training (BMET) under the Ministry of Labour and Manpower. The Program covers 67 *thanas* in 51 districts and the beneficiaries include rural landless poor men and women with land of up to 0.5 acre. A sum of Taka 99.4 million has so far been disbursed to 26,000 beneficiaries and the rate of recovery is about 56%. Efforts are under way to improve the delivery, monitoring and recovery system of the overall credit program.

The **Rangpur Region Rural Development Project (RD-9)**, financed by the EC, is being implemented in greater Rangpur district. Although the Project is being financed under the aegis of Bangladesh Rural Development Board, the uniqueness of the intervention lies in the fact that the loans are being channelled through informal groups and the funding is being made jointly by the donors and the Sonali Bank, with provision for joint risk sharing. The bank has so far disbursed Taka 194.7 million to 65,000 beneficiaries with a recovery rate of around 100%.

The **Mahila Rin Karmashuchi** (Women's Credit Program) of the Bank implemented in 130 *thana* of the country under the aegis of the Bangladesh Rural Development Board is yet another remarkable intervention in poverty alleviation through the creation of gainful employment opportunities, exclusively for women. This is probably the largest of the credit programs run exclusively for and by women in Bangladesh. The Bank has so far disbursed a sum of Taka 471.1 million to 0.15 million beneficiaries and the average recovery is around 91%. Here the uniqueness lies in the fact that the program is being run through a cooperative structure and it has so far been very successful.

Almost all the poverty-focused micro-credit programs described above have the following common features:
- loans are given without any collateral security
- loans are disbursed to the individual through homogeneous groups and also through the *bittahin samabaya samity*
- group members stand guarantee for each others' loans
- the borrower has to make some mandatory savings
- the loans are intensively supervised and closely monitored
- the major beneficiaries include rural landless poor men and women with land up to maximum 0.4-0.5 acre and wage labour as their principal means of income
- each group comprises 5 to 20 members and the loan ceiling for an individual is Taka 5000 to 7000; the mode of repayment is mostly in weekly instalments
- loans are disbursed for any legal income-generating activity: farm and off-farm productive investment; small scale dairy (including beef fattening, milking cow); poultry and fishery projects; crop production; and cottage industry etc.
- The bank has so far disbursed a sum of Taka 2058.8 million under various poverty alleviation programs to about 0.68 million beneficiaries, out of which about 0.5 million are women.

Agrani Bank

The Agrani Bank's exclusive credit program targeting poverty alleviation includes the following:
- Productive Employment Project (PEP) run in collaboration with SIDA & NORAD
- Daridra Bimochon Karmasuchi (DABIK) run on the Bank's own initiative and resources.
- IFAD assisted Small Enterprises Development Project.
- DANIDA assisted Noakhali Integrated Rural Development Program (NIRDP)
- Agrani Bank—BRDB collaboration on Rural Poor Program.

The **Productive Employment Project (PEP)** was implemented by the Agrani Bank in collaboration with the Bangladesh Rural Development Board (BRDB) with financial assistance from the Swedish International Development Agency (SIDA) and the Norwegian Agency for Development Cooperation (NORAD). The project was originally launched in six *thanas* of Faridpur and Kurigram district in 1986-87 with the objective of raising the socio-economic condition of the rural poor through the provision of skill development and awareness-raising training, followed by credit for undertaking suitable income-generating activities. Following success in both raising the income level of the target group and recovery of credit, the project area has gradually been expanded and its current phase is being implemented in 30 *thanas* of Faridpur and Kurigram districts. So far, the Project has enrolled 56,415 rural poor members out of which 51,840 have received credit of Taka 17.63 crore from Agrani Bank. the poor members of PEP have made a thrift savings of Taka 2.56 *crore* within this period. The rate of recovery of credit is 99%.

Janata Bank

The Bank's exclusive credit programs include the Small Farmers and Landless Labourers Development Project, the Diversified Credit Program and Rural Poor Program. The Janata Bank and the Bangladesh Academy for Rural Development (BARD) have jointly been implementing the Small Farmers and Landless Labourers Development Project (SFDP) in 51 *thanas* of Comilla, Bogra, Mymensingh, Patuakhali, Borguna, Barisal and Bhola districts since 1988-89. This is an experimental project run with some financial support from the UNCDF. The immediate objectives of the project are: to increase production, employment and income of small farmers and landless labourers through the formation of small homogeneous groups; the inculcation of a savings habit among group members in order to generate their own capital; and provision of credit support to them through a commercial bank to allow them to under take various Income Generating Activities (IGAs). So far, the Project has enrolled 70,914 rural poor members who have received credit of Taka 259.33 million from the Janata Bank. The rate of recovery of credit is 99%.

The Bangladesh Krishi Bank (BKB)

The BKB's exclusive poverty alleviation credit programs include the following :
- BKB-UNCDF collaborative credit program
- BKB-ESCAP/ILO collaborative credit program
- BKB-BSCIC collaborative program for rural women
- ADB assisted Rural Women Employment Creation (RWEC) Project
- Employment Program for Educated and Trained Youth
- Marginal & Landless Farmers Credit Program.

The Marginal and Landless Farmers Credit Program was been launched in June 1995 by BKB involving all branches to provide collateral-free credit to the rural poor for undertaking various IGAs. The ADB-assisted Rural Women Employment Creation Project has been implemented in 12 *thanas* of the country since 1993. The project aims at creating employment opportunities for target beneficiaries through the intervention of local NGOs.

Issues and Recommendations

Linkage with NGOs & the Rate of Interest

A close examination of the performance of the above projects in disbursement and recovery of poverty-alleviation credit reveals that the projects have achieved, if not out-performed, the level of recovery of the NGOs. It belies the generalized statement and belief that the banks are not inclined to disburse credit to the rural poor without collateral security. The educative revelation on this score is that it is not only the institution which accounts for better recovery, it is also the area of intervention and the type of clientele, along with the mechanism of delivery and monitoring that results in good performance. The banks have a large number of branches in far-flung areas; they have surplus investment funds with ample potential to meet the higher credit needs of 'graduated' beneficiaries, whatever their needs. Since the empirical evidence of the performance of the projects detailed above shows that the banks can undertake NGO-type activities and achieve a satisfactory level of performance, the time has come for all the banks to embark upon similar poverty-focused credit programs on a large scale, involving all their outlying branches and surplus funds. Since the banks are not constrained by funds, the human capital input commonly provided by donor-assisted projects such as those described above could be arranged, thus forging effective relationships with the scores of NGOs rich in experience but lacking material resources.

There are two possible models for such a relationship. In the first case, NGOs could take bulk loans to pass on to grass-roots beneficiaries. The Agrani Bank's link with ASA follows this model. The second option is banks lending directly to beneficiaries with NGOs intervening at the social preparation stage in lieu of service charge, offering: group formation; motivation; training; supervision etc. The IFAD/GTZ-assisted MSFSCIP tested this model in Kurigram. Most of the NGOs prefer the first option i.e., taking bulk finance for onward retailing. I am inclined to recommend the second one on the following grounds:

- bulk lending would call for collateral security because of the amount of money; the Agrani-ASA linkage was delayed on this issue.
- bulk lending does not help the beneficiaries and the banks to establish a permanent relationship with each other.
- bulk lending would tend to turn the financing bank into a sleeping partner; the banks would never develop institutional expertise in handling poverty-alleviation credit programs.

One may raise here the question of the cost of intervention in the case of the second option. This can be resolved by a rationalization of the rate of interest at grassroots level. For instance, the NGOs and donor assisted projects charge interest at around 20% for poverty-focused credit programs, and various studies on the issue have shown that it is not the rate of interest, rather the timely availability and adequacy of the credit which matters most to the beneficiaries. Accepting this statement, the rate of interest at grassroots level may be fixed at 20%. As the cost to the banks is somewhere below 8%, they could charge interest at a rate of 10-12% (as they do in their own poverty-focused credit programs), adding to this a service charge of 8-10% to cover the intervening agency's (the NGO's) costs.

Management Issues

With the linkage and interest issue taken care of, I will go on to mention some other management issues concerning improvements in credit service to the rural poor.

Graduated Members and the Scale of Finance

The banks have now become accustomed to lending without collateral security in the case of micro-investment for poverty alleviation. Physical collateral has been replaced by group security. The banks are well placed because of their resource base and their ability to adjust the size of loan for 'graduated' members of the group. Unlike NGOs, banks can easily meet the higher credit needs of graduated members with technologically upgraded enterprises. The banks should, therefore, launch feeder programs for graduated members.

Supervision and Monitoring

Empirical evidence dictates that proper supervision and objective-oriented monitoring is the magic which achieves targeted recovery in poverty alleviation credit programs. The Sonali Bank, for instance, launched a Special Investment Scheme in 1993 with a view to providing credit on easy terms to small and potential entrepreneurs wishing to undertake small scale dairy, fishery and poultry projects. The Scheme has provision for giving loans up to Taka 50,000 without any collateral security. The bank has, so far, disbursed over Taka 550 million to more than 10,000 small entrepreneurs. The Scheme is maintaining a loan recovery level of over 80%. The magic ingredient has been that the bank has made a little bit of extra investment in the development of intensive supervision and monitoring capabilities, exclusively for this scheme.

Motivation and Training

There is no denying the fact that micro-credit management calls for the selfless services of well-trained and motivated personnel. Bank employees serving in rural areas in a hostile atmosphere would do their utmost to get an urban-based posting for obvious reasons, unless there is any extra incentive for them. This is true for all the rural branches of all banks handling rural credit in general and poverty-focused micro-credit in particular. Special training and an incentive package is, therefore, a *sine qua non* for effective management of micro-credit in rural areas.

Policy Commitment and Enforcement

Almost all the poverty-focused credit programs with good recovery records handled by the banks have donor backing. It has been observed that when donor support is withdrawn, activities slow down. One of the reasons for the decline in project performance after the donor's withdrawal is the lack of commitment from the mainstream management of the banks. After the donor's withdrawal, projects are swallowed up in the bank's general credit program and thus the focus is lost. The banks' top management needs to make an explicit commitment to poverty alleviation programs, and the banks' annual lending programs should have specific targets for disbursement in such programs.

Concluding Remarks

Banks have been being condemned time and again in various discussion forums/seminars for their inability to meet the national need to increase efforts directed at poverty alleviation. Such observations are not based on facts as is evident from the above discussion. The banks have a clear and ambitious program for lending in the rural sector. But they also have problems, which need to be looked at. Their efforts should be directed towards improving credit delivery and their monitoring mechanism.

There is a growing tendency among some experts to propose the establishment of new conduits for channelling micro-credit, leaving the 'problematic' existing institutions to one side. This is an expedient solution, since addressing problems within the existing institutions is a more than Herculean task. But Bangladesh, as a resource-poor nation, should think twice before agreeing to such experiments. There have been a score of Rural Development Projects with a credit component for poverty alleviation implemented since Independence, involving the banks in dispensing micro-credit. But none of the projects addressed the institutional deficiencies of the Participating Credit Institution (PCI). The result has been that projects have folded and the Bangladesh Bank has withdrawn the temporary refinance facility allowed to the PCI against project lending. Everyone involved in the process has washed their hands of it, but the banks are left having to use depositors' money to carry the burden of huge, overdue project loans. That a little institutional development support to the lending banks can do magic has been proved by the performance of the SIDA/NORAD-assisted Productive Employment Project of the Agrani Bank, and the ADB-assisted Rural Poor Cooperative Project of the Sonali Bank. The former targets beneficiaries through informal pre-cooperative groups and the latter targets beneficiaries through the cooperative structure. The projects' success in maintaining a 100% recovery rate belies the common notion that formal banks as well the cooperative sector are not capable of handling poverty alleviation credit programs efficiently. The problem lies elsewhere, and demands the thorough scrutiny of the experts.

Annex 1
SWANIRVAR CREDIT PROGRAMME

Union : 1024
Thana : 135
District : 39

Month: April '96

Sl No	N● OF PARTICULAR	SONALI	B K B	AGRANI	JANATA	RUPALI	R K G B	PUBALI	UTTARA	TOTAL
1	No o" Bank Branch	156	183	125	102	47	54	22	1	700
2	No of Unions	243	256	174	146	69	133	32	1	1024
3	No of Village	2593	2466	1625	1712	778	736	447	7	10429
4	No of Groups	32274	35010	24042	19870	8519	9975	3990	74	133555
5	No of Loanees	158386	168031	121267	93828	42928	43011	19450	39	645940
6	No of Project	16	47	12	20	18	18	12	1	47
7	Amount Disbursed	325670517	446552450	243617623	143537900	86170200	100275261	25109700	35100	1370972751
8	Due for Recovery	623880082	585378723	366494415	211809107	144902710	142657399	43669600	59334	2119857270
9	Amount Recovered	302224870	398814312	226281974	128832688	83326762	78432355	29990372	28142	1247931985
10	Rate of Recovery (Including Interest)	48.36	68.13	61.74	60.82	57.51	54.98	65.68	47.43	58.87

Annex 2
POSITION OF MSFSCIP CREDIT OPERATION

PERIOD: Up to 30-06-96

NAME OF BANK	TYPE OF GROUP	TOTAL NO OF GROUP ACCOUNTS			NO OF LOANEE GROUPS			DISBURSED	LOAN STATUS				RATE OF RECOVERY
		M	F	TOTAL	M	F	TOTAL		RECOVERED	OVERDUE	OUTSTANDING		
RAKUB	MARGINAL	667	638	1305	585	508	1093	779.51	535.74	22.17	342.87		96.0%
	SMALL	391	40	431	215	2	217	181.67	114.93	1.78	73.10		98.5%
AGRANI	MARGINAL	102	117	219	118	95	213	120.53	76.18	7.84	54.23		90.7%
	SMALL	213	17	230	134	3	137	119.36	73.08	0.93	55.17		98.7%
SONALI	MARGINAL	92	105	197	79	99	178	82.68	57.83	1.72	31.86		97.1%
	SMALL	147	13	160	97	4	101	82.05	48.95	0.74	36.50		98.5%
JANATA	MARGINAL	79	75	154	81	64	145	124.80	87.33	1.89	48.76		97.9%
	SMALL	52	2	54	25	0	25	25.17	12.98	0.19	13.31		98.6%
TOTAL	MARGINAL	940	935	1875	863	766	1629	1107.52	756.58	33.62	477.72		95.7%
	SMALL	803	72	875	471	9	480	408.25	249.94	3.64	178.08		98.6%
	SMALL	1743	1007	2750	1334	775	2109	1515.77	1006.52	37.26	655.80		96.4%

Chapter 9

GRAMEEN BANK: A CASE STUDY

Syed M Hashemi, Lamiya Morshed

Introduction

Grameen Bank began as an experimental project in rural Chittagong in 1976. The project was initiated by Muhammad Yunus, a university professor at Chittagong University in eastern Bangladesh to provide affordable credit to the poor as a means of combating poverty in rural Bangladesh. Yunus had observed that the poorest of the poor had no access to commercial credit and were therefore being systematically exploited by moneylenders in the traditional money markets, which further exacerbated their poverty. The project began by providing small collateral free loans to the rural poor for income generating activities chosen by the borrowers themselves. The borrowers were required to form groups of five persons which served as a screening and monitoring mechanism replacing the need for collateral and ensuring that transactions costs would below.

Yunus recognized that rural women were particularly vulnerable because of their restricted access to and control over resources and the project which had begun lending to both and men women, moved gradually towards lending primarily women as women demonstrated that they invested more carefully and repaid more faithfully.

Following the experience of high repayment rates, the project expanded over the next years into neighbouring districts and in 1983 became established as a specialized financial institution under a separate government statute in 1983 with a mandate to lend to the poor. The Bank is governed by a 13-member board of whom 9 are Grameen bank borrowers and three are government appointed senior civil servants. Ninety two percent of Grameen Bank shares are held by members themselves while the rest are held by the government of Bangladesh.

Funding for Grameen Bank has been primarily in the form of loans with more than three quarters coming from the Bangladesh Central Bank and local commercial banks and the rest from international donors. Funding for the Grameen Bank had been at preferential rates in the past but since last few years, Grameen has been borrowing from local money markets at commercial rates of interest.

Grameen Bank charges interest rates close to those charged by commercial banks, which compare favourably with the usurious rates charged in the informal credit markets in rural areas. This has enabled the Bank to attain operational self sufficiency and a World Bank study published in 1995 shows that the Bank is moving towards full self sustainability.

The Bank today lends to over two million borrowers, 94% of whom are women, and with repayment rates as high as 98 percent, GB has shown that not only are the poor, particularly poor women, bankable but that lending to the poor can be far less risky than lending to the rich. The Grameen Bank has now become a model for group based lending to the poorest throughout the world.

Explanations of Effectiveness

The effectiveness of the Grameen system is attributed to a number of interrelated features. Grameen Bank targets a well defined homogeneous clientele. This is the large and fast growing population of landless rural poor, where landless is defines as someone from a household that owns less than 0.5 acres of cultivable land or assets with a value equivalent to less than 1.0 acre of medium quality land. Grameen Bank borrowers are expected to select from among their peer group to form a group of five women. This self selection screens out the non-poor and those persons that would not be able to make loan repayments. The Grameen system operates in a social milieu where women are willing and able to pressure one another to repay loans. Women who doubt that they will be able to make timely payments and otherwise comply with the rules and requirements of Grameen Bank are reluctant to join. The small size of loans, and the identification of the Bank with the poorest, discourage women from wealthier families from joining. As members of GB are expected to deviate from traditional norms, attending weekly meetings in public places and interacting with male staff, the system ensures that those who have nothing to lose are those who join. Before the group is recognized by the Bank, the members go through a seven day long training during which they learn the rules and regulations of the program, learn to sign their names and learn the sixteen decisions. The groups of five are organized into centers with each center consisting of six to eight groups. Each group elects a group chairman and group secretary who are responsible for ensuring attendance at center meetings and maintaining discipline during the meetings.

Members are eligible for a range of different loans. In the first year, members receive the general loan which has a repayment period of one year and ranges from US$ 75 to US$ 100. The loans are disbursed to the two neediest members of the group who must repay on schedule before the next two receive the loan with the Group Chairman of the group receiving the last loan. The borrowers who are awaiting loans will try to ensure that repayments are made on time. A borrower will make a loan proposal to the Bank. The loan can be used for any income generating activity chosen by the borrower herself but approved by the group members and the center chief. The loans are repaid in 50 equal installments and the interest payment is 10 percent of each principal installment. The small installments make it easier for borrowers to pay. When a loan is repaid in full, the borrower will be entitled to another, often larger loan. This works as an incentive for borrowers to repay. If a borrower willfully defaults, her group members will pressure her to pay as they are liable for unpaid installments. Borrowers know each other and are in a position to ensure repayment. If a borrower are unable to repay, then the other borrowers will either provide the support by paying an installment for her. This system of joint liability and collective responsibility mechanisms contribute to high repayment rates.

All borrowers are expected to save. Savings behaviour begins during the training period when borrowers save one taka per week into a joint account. A compulsory deduction of 5% of the loan is made at the time of disbursement of the first loan for the group fund. The group fund also includes a personal savings deposit of 1 taka per week by each member. When the amount of savings deposited in a Group Fund reaches 600 taka, the group of five is obliged to purchase Grameen Bank shares in the amount of 500 taka at the rate of 100 taka each. Savings provide security against default, is an economic buffer for the most vulnerable clients and becomes a source of additional loans, for investment or consumption, the terms and conditions of which are determined by the group members themselves. Managing the group fund helps to create unity within the group and gives the members experience in the collective management of assets. In addition, borrowers are encouraged to make voluntary deposits into individual accounts with Grameen Bank which provides interest at the rate of 8.5 percent per annum. Centers meet weekly and attendance by all members of the center is mandatory and yet another mechanism whereby individuals who lack the sincerity and discipline to repay their loans on schedule are excluded. All banking transactions take place during these meetings including collection of loan repayments and savings. These transactions are conducted at the group meetings to ensure transparency.

Access to bigger and different types of loans depends on timely repayment of the early loans and acts as an incentive to repay. In addition to the general loan, borrowers may apply for seasonal loans, housing loans, technology loans, crop processing loans and other seasonal activities. Members are eligible for a range of different loans with varying rates of interest and varying repayment periods. The housing loan for example is a long term loan that requires weekly repayments over ten years at a rate of 8 percent per annum.

The rules, rituals and procedures of the Grameen Bank appear to have multiple functions. They provide an efficient way to conduct financial transactions, provide a screening mechanism, inspire a sense of loyalty and responsibility to Grameen Bank, and promote self confidence among the female borrowers. The working style of Grameen Bank takes into account the characteristics, constraints and social norms of its clients. Financial transactions are conducted in the village rather than at the branch office, making it accessible to women, who rarely leave the village and who would most likely be intimidated by the atmosphere of a commercial bank. The rules and procedures are simple and readily understood by the clients and do not require that the members be literate. The simplicity of the systems facilitates the self screening of potential participants, to a large extent they know what they will be getting into. Living in small, relatively traditional communities, the GB members and prospective members have access to the information needed to evaluate the credit worthiness of their neighbours.

Organization Structure

The Grameen Bank is organized at four administrative levels: the branch, the area, the zone and the head office. The lowest administrative unit is the branch which employs a staff of ten people. Each branch serves approximately 50 to 60 groups of 30-40 borrowers. The branch is established in a rural area where surveys show are inhabited by large numbers of poor people who fall within the target group. The branch is supervised by an area office which is staffed by six people. Each area office covers 10-15 branches. The area office in turn is supervised through 14 zonal offices. Each zonal office has a staff of 35.

The branches maintain day to day contact with the clients and the area offices are directly managed by the 14 zonal offices. The zonal offices are independent to the point where even strategic decisions can be taken by them. Although a system has developed over the last two decades, at every level, Grameen Bank operates on the premise that it is a system that can continue to improve. For this reason, experimentation is encouraged at the zonal level without clearance at the Head Office. Staff are encouraged to 'learn by doing'. They are encouraged to develop creative solutions and bring forward their own personal views and ideas and criticisms.

Grameen Bank currently has a staff of over 12,000 employees. All staff undergo an intensive 12 month long training program. The first six months is the "induction training period" which is held at the head office at the completion of which the trainees sit for an examination to proceed to the second stage of their training during which each trainee is posted in a Grameen Bank branch during which time they assume the responsibilities of a bank worker. The trainees become permanent employees after passing one more exam. This core training is supplemented by additional training in computerization, accounts, administration, leadership and crisis management.

Grameen Bank has a comprehensive management information system, paper based at the field level and computerized at the central level. Through all levels of the management and personnel there is a flow of information and comprehensive data accumulation and report preparation reflects the Bank's attempt to set in motion a learning process. Like the major credit delivery NGOs in Bangladesh, the Grameen Bank gathers socioeconomic data to monitor and evaluate the viability and impact of credit to the landless.

Performance

Grameen Bank has increased its a membership of 2,064,011, 94% of whom are women in 35,697 villages, over half of the villages in Bangladesh. Studies of Grameen Bank (Hossain, 1988; World Bank, 1995; Pitt and Khandker, 1994) confirm that the Grameen Bank is succeeding in reaching its target clientele. Over US$ 1.5 billion in loans had been disbursed by January 1996. Housing loans amounting to US$ 135 million have been disbursed to build 330,000 low cost, sturdy houses by December 1995. By January 1996, GB members had generated a cumulative savings of 125 million dollars, which is far more than that mobilized by the country's commercial banks.

Grameen Bank's assets have grown steadily from 119 million taka in 1983 to 16.6 billion taka in 1994. Grameen Bank's operational expenses totaled about US$ 50 million in 1994. Operational expenses are primarily accounted for by salaries, interest paid on deposits, interest on loans as provisions for bad debt. In 1994, Grameen generated a total revenue of about $50.5 million to earn a profit of a little over half a million dollars. Eighty-two percent of revenues were generated from interest on loans made out to members. Revenue also includes interest from fixed deposits and three month government treasury bills and four percent from a grant from NORAD for head office expenses. The share of grants from foreign donors as a percentage of income has declined from 20% in 1989 to 3.8 percent in 1994.

The basic profit making unit is therefore the branch which borrowers funds from the head office at 12% and on-lend at a 20 % to borrowers. The 8% margin represents the largest source of earnings in addition to earnings from deposits of group funds and other savings. Branches become profitable after their fourth or fifth year of operation when disbursement exceeds ten to eleven million taka and membership reaches 1600 to 1700 poor borrowers. A study by Hashemi and Schuler has shown that a branch can become financially self sufficient and viable after five years of operation after reaching a level of disbursement of around 12 million taka with a fund requirement to take the branch from start up to sustainability of 6.3 million taka. Although Grameen Bank is strictly speaking being subsidized, it currently makes a profit and its subsidy dependence has been declining.

Economic Impact

The economic impact of the Grameen Bank has been examined by two major studies, one by Hossain in 1988 and the other by Khandker and Chowdhury in 1995.

Hossain's study was based on detailed household information from a random sample of 200 households in five villages where branches were over three years old and from 80 households in two control villages indicated a positive impact on the economic status of Grameen bank borrowers. Loans had generated new employment for a third of the sample, who reported that they were unemployed before joining the Bank. The average level of employment for members increased from about six working days to 18 working days per month. Incomes in member households were seen to be 43% higher than in target group households in control villages and 28 percent higher than in non-participating households in the Grameen villages. These higher incomes were attributed by the study to increases in income from the processing and manufacturing, trading and transport services financed with loans from Grameen Bank. The study also showed increases in per capital food consumption in member households, and more investments in housing, education and sanitation.

The second study which was conducted by the World Bank in collaboration with the Bangladesh Institute of Development Studies used data on consumption, savings, asset ownership, and net worth to assess the economic impact of microcredit programs. Data was collected as part of an extensive survey of 1798 rural households which included Grameen Bank members and members of two other credit programs and groups of comparison households from credit program and non-programme villages. The researchers findings showed that Grameen Bank and to a lesser extent the other two programs studied reduced poverty and improved welfare of participating households but also enhanced the household's capacity to sustain their gains over time. This was accompanied by an increased caloric intake and better nutritional status of children in households of Grameen Bank participants. The study also found evidence of a shift from traditional farm activities towards non farm activities which are the main potential sources of productivity and growth in rural economy such as Bangladesh's.

Social Impact

Observers of the Grameen Bank believe that it is bringing about fundamental changes in the lives of the women it serves, improving their self perceptions, their positions within the family and community. The women go against traditional norms, meeting weekly in public place, interacting with men outside the community. They become, in most cases for the first time in their lives, involved in financial transactions and are encouraged and enabled to conduct independent economic activities and acquire assets in their own names. A study by Hashemi and Schuler showed a significant ($p < .05$) effect of membership in Grameen Bank and other credit programs on women's empowerment and on contraception use. The study analysed the effects of GB women's contribution to family support and a composite empowerment indicator (which aggregates eight different indicators of women's empowerment, including, among others, physical mobility, economic security, participation in important family decisions) Duration of membership had a significant effect on both women's contribution and the composite empowerment measure. The study on reproductive behaviour also showed that duration of participation in a credit program had a significant effect on current contraceptive use.

Conclusion

The greatest obstacle that Grameen faces in providing targeted credit for poverty alleviation is poverty itself. Economic stagnation in rural areas limits the potential for productive investments. Lack of other social services for the poor in rural areas leaves Grameen borrowers highly vulnerable to external shocks. Grameen continues to develop new initiatives to address these problems. These are funded separately from the credit program and could threaten the organization's sustainability; but this awareness has not led GB to subordinate the goal of poverty alleviation to that of sustainability.

Table 1: Growth in Operations

(Amounts in millions of taka)

	1983	1984	1985	1986	1987	1988	1989	1990	1991	1992	1993	1994
Total no. of branches	86	152	226	295	369	501	641	781	915	1015	1040	1045
Total no. of villages	1249	2268	3666	5170	7502	10552	15073	19536	25248	30619	33667	34913
Total no. of members	58320	121051	171622	234343	339156	490363	662263	869538	1066426	1424395	1814916	2013130
Total no. of loanees	46995	106943	152463	209467	328557	472430	648267	852622	955031	1385324	1682914	1860674
Total loan disbursement	195	499	928	1470	2280	3560	5328	7591	10230	15434	26056	39968
Total housing loan disbursement	—	4	21	27	167	338	574	799	1100	1660	3333	4671
Total member savings (Group fund)	16	38	71	115	168	297	451	650	892	1308	2117	3147
Total income	1	36	66	90	129	200	299	407	532	773	1325	2019
Total expenditure	4	31	65	90	128	198	297	404	540	778	1316	1997
Net profit for the year	-3.1	4.9	0.5	0.4	0.4	1.2	2.3	3.1	-8.3	-5.7	9.6	21.7

Table 2: Balance Sheet

(in millions of taka)

	1983	1984	1985	1986	1987	1988	1989	1990	1991	1992	1993	1994
Property and Assets												
Cash in hand	0.1	0.0	0.0	0.0	0.3	0.0	0.5	0.0	0.0	0.0	0.1	0.1
Balance with other banks	12.7	20.0	19.1	19.4	22.3	34.6	49.1	60.1	67.1	151.7	339.9	328.9
Investment	26.5	146.5	218.5	435.5	408.3	399.3	642.0	1077.7	1450.8	1301.2	1744.8	3201.9
Loans and advances	74.3	177.5	245.7	331.0	644.1	1094.9	1593.2	2117.4	2551.2	4423.9	8763.6	11053.0
Fixed assets	1.4	4.8	9.7	32.1	68.5	122.6	161.8	264.3	344.0	421.1	497.1	551.2
Other assets	3.9	27.9	53.0	82.0	135.9	259.7	436.1	581.0	427.0	585.7	931.1	1439.2
Total	118.8	376.6	546.0	900.0	1279.3	1911.3	2882.7	4100.4	4840.1	6883.7	12276.6	16574.1
Capital and Liabilities												
Capital–paid up	18.0	27.2	30.0	35.5	42.1	56.9	72.0	72.0	114.4	149.5	150.0	216.5
General and other reserves	–	1.0	1.5	1.8	2.7	3.7	5.8	8.7	8.7	21.4	49.9	70.4
Revolving funds	-3.1	0.6	0.5	0.5	7.7	10.3	307.6	1025.0	1278.1	2383.5	2828.8	3089.2
Borrowings from banks/institutions	85.4	311.3	433.9	716.9	954.7	1232.9	1720.0	1838.2	1876.1	1878.2	5470.0	8215.7
Deposits and other funds	18.5	38.3	79.8	145.0	221.9	324.5	566.7	851.4	1381.0	2176.3	3150.2	4132.6
Other liabilities	0.0	0.2	0.3	0.4	38.2	215.9	91.6	168.7	183.0	274.8	628.8	849.7
SIDE loan payable Acct. (Contra)	–	–	–	–	12.0	67.2	119.0	136.9				
Total	118.8	376.6	546.0	900.0	1279.3	1911.3	2882.7	4100.4	4841.3	6883.7	12276.6	16574.1

Table 3: Profit and Loss Account

(in millions of taka)

	1983	1984	1985	1986	1987	1988	1989	1990	1991	1992	1993	1994
Income												
Interest on												
Loans and advances	0.2	23.4	35.1	42.6	65.1	112.7	162.0	213.7	315.7	522.2	1055.6	1646.4
Investments	0.0	0.1	0.8	1.6	59.7	51.6	74.0	112.9	133.6	176.4	165.6	262.1
Deposits	0.7	11.9	29.3	45.7	0.5	0.8	1.1	1.6	1.6	2.0	3.7	4.4
Other Income	0.2	0.7	0.1	0.3	3.5	34.5	62.0	78.6	80.9	71.8	100.3	105.6
Total	1.1	36.0	65.7	90.2	128.7	199.6	299.0	406.8	531.8	772.9	1325.2	2018.5
Expenses												
Interest on												
Deposits	0.3	1.8	4.	0.8	12.5	20.9	32.8	54.6	68.8	101.9	151.9	269.4
Borrowings	1.1	14.1	25.5	23.6	22.2	27.7	38.7	43.2	50.3	89.0	235.5	522.5
Salaries and other related expenses	1.3	9.4	25.4	43.7	65.3	81.1	121.2	170.8	268.6	399.2	579.8	626.6
Directors' remuneration	0.0	0.0	0.0	0.0	0.0	0.0	0.0	0.0	0.0	0.0	0.0	0.0
Other expenses	1.5	5.0	9.0	13.7	26.3	65.9	69.1	124.0	135.4	178.0	330.3	558.0
Depreciation	0.1	0.7	1.1	1.4	1.9	2.9	8.0	11.1	16.9	19.9	18.2	20.3
Total	4.2	31.1	65.3	89.9	128.3	198.4	296.8	403.7	540.1	778.1	1315.7	1996.8
Net profit for the year	-3.1	4.9	0.5	0.4	0.4	1.2	2.3	3.1	-8.3	-5.7	9.6	21.7

PART III

PROBLEMS OF REACHING THE POOREST

Chapter 10

MICRO-CREDIT PROGRAMS: WHO PARTICIPATES AND TO WHAT EXTENT?

Hassan Zaman

Introduction

Land ceilings, occupational criteria and asset valuations are commonly used for targeting purposes by credit agencies aiming to direct resources to the rural poor. However a mixture of demand and supply side factors leads to the inclusion of a small group of 'non target' households in these credit programs. This paper starts by examining the differing characteristics of the 'properly targeted' versus these 'non eligible'[1] members. The next section uses multivariate analysis to identify the characteristics which lead to participation in a credit program. The second part of the paper looks at the 'depth' of participation of program members using a set of credit based indicators. Differences between 'correctly targeted' and 'non target' households are examined in terms of 'participation depth' and multivariate analysis is again used to shed light on the possible determinants of active participation. The concluding section looks at the implications of the earlier analysis for micro-credit policy. The paper uses data collected by the author as a team member of the BRAC-ICDDR,B joint research project in fourteen villages in Matlab *thana*[2], Bangladesh. BRAC's Rural Development Program (RDP) has been

---

[1] The term 'target group' (TG) will be used interchangeably with 'eligibility' in this paper.
[2] Matlab is a sub-district in Chandpur district 55 kms southeast of the capital Dhaka with a population of about 400,000.

operating in ten of these villages for three years and will be used as the micro-credit program under study.

Initial Endowment: Are BRAC Member's 'Homogenous'?

A comprehensive village survey in the fourteen villages found a total of 585 BRAC members and 2935 non members. 72% of the BRAC members were classified as 'target group/eligible' (TG) and 28% as non-target group (NTG) by the field investigators using BRAC's official targeting criterion.[3] One must note that since the research was conducted three years after BRAC's RDP started operating in Matlab some of the households classified as NTG could have 'graduated' from TG to NTG status in the interim.[4] Table 1, based on data collected in these fourteen villages, suggests that the BRAC member NTG households are on the whole considerably better off compared to TG households in almost all indicators of well being. However this NTG group are less well off using the same indicators when compared to the non BRAC NTG group (see Table 1). In other words, the members who fall outside BRAC's official targeting criterion but are 'mistakenly included' come from a category of households that could be considered better off compared to even the 'marginal poor'[5] category but not to the extent that they form part of the village elite. The tests of differences in means and proportions showed that there is little to choose between TG member households and TG non member households in certain dimensions (average education, total savings, dependency ratios, remittances received, proportion of manual labourers, 'hunger', food share) whilst considerable differences exist in others. For instance, target group non members seem wealthier in that they have significantly more land and greater value of non land assets compared to target group members and also they have a larger ratio of earners to household members. TG member households on the other hand have significantly

---

[3] BRAC targets households whose land ownership is less than 0.5 acres (50 decimals) and whose main source of livelihood is manual labour (this criterion constitutes 'BRAC eligibility').

[4] Moreover the figure for Matlab is higher than a similar calculation from a nationally representative study which found that the non target proportion in BRAC groups is 20% (Mustafa/Ara 1995) possibly due to the rapid scaling up of RDP the year the Matlab branch was opened.

[5] See Rahman and Hossain (1995) for the need to include the 'marginal poor' in targeted anti-poverty programs.

fewer heads of household who were ill in the last fifteen days, younger household heads and a greater proportion of male headed households.

Table 1: Summary Statistics for Different Socio-Economic Groups in the Sampled Villages

Indicators	BRAC Member TG n = 421	BRAC Member NTG n = 164	Non BRAC Member TG n = 1569	Non BRAC Member NTG n = 1366
Quantity of land owned (decimals)	13.70	87.91	17.13	134.22
Value of non land assets (taka)	15943.00	44159.00	15828.00	83849.00
Food share in total consumption	0.71	0.67	0.73	0.64
Proportion of HH's which went without rice/chapati for one day in last four months	0.06	0.0	0.08	0.01
Daily consumption per capita (all items) in taka	14.40	20.87	14.07	31.43
Total credit taken in last four months (taka)	2665.00	3234.00	2053.00	4057.00
Total savings (in taka)	2455.00	8292.00	2226.00	9933.00
Proportion of HH heads who are manual labourers	0.33	0.07	0.31	0.04
Proportion of HH heads ill in last two weeks	0.14	0.12	0.16	0.10
Proportion of HH's seriously damaged in last four months	0.07	0.05	0.06	0.03
Average years of education in household	1.31	2.69	1.40	3.69
Age of household head	42.70	46.52	44.44	51.23
Dependency ratio*	0.33	0.29	0.34	0.31
Ratio of earners to total members	0.22	0.20	0.28	0.23
Remittance received last year in taka	162.00	358.00	82.00	804.00
Proportion of male headed households	0.87	0.85	0.82	0.80

Note : *(number under nine + number over 60)/(number between 10-60).

Source : The data was collected by the BRAC-ICDDR,B Matlab project between April-August 1995.

A simple demarcation between the poor and the ultra poor (see Lipton 1983) can be made using the land ownership data. Nearly half of BRAC members have less than ten decimals of land (47%) and 30% have less than five decimals. The national rural proportion of households with less than five decimals is 17.6% (BBS 1995) thereby suggesting that BRAC groups have more than a proportionate share of ultra-poor households.

Hence two indicators that are strongly correlated with poverty namely landlessness and female headship, give mixed messages in terms of the accessibility of micro-credit programs for the poorest. The land data suggests the ultra poor do take part while the relative lack of participation of female-headed households suggests that there may be barriers to entry for the most vulnerable in society.

Having established that basic differences exist among BRAC members the issue of whether other competitive forces in the village affect the socio-economic profile of the type of household selected by BRAC was looked into. The one village out of the ten in the sample which had BRAC, ASA, BRDB and the Grameen Bank[6] was compared with villages which had either only BRAC or BRAC plus one other organization. Tests of differences in means were carried out between the two sets of villages, e.g. the land owned by the typical BRAC target group member in the 'competitive' village versus the mean land owned by a BRAC target group member in the 'non competitive' village. The tentative finding is that the degree of 'competition' does not make much difference to the 'type' of household participating in BRAC although the small number of villages under study make generalizations difficult. The differences in 'endowments' that do occur between households in the two sets of villages are more likely to be due to village level differences as when differences do occur they are consistent amongst the different socio-economic classes. The village where program concentration is the highest appears more prosperous (measured by a number of indicators) than the norm.

The Determinants of BRAC Membership: A Multivariate Analysis

'Participation' in a targeted credit program is the outcome of both demand-led and supply-side factors. The former depends on the

---

[6] ASA (Association for Social Advancement), BRDB (Bangladesh Rural Development Board) and the Grameen Bank are other agencies involved in lending to the rural poor.

judgments of eligible households about the costs and benefits of taking part; the supply aspect revolves around the decision by the organization to locate in a particular village and secondly to select households for the program.

This section uses multivariate analysis to assess the factors influencing a woman's decision to join, or be selected by, BRAC. In order to do so one must identify certain explanatory variables that can be considered exogenous and hence exclude others such as savings, non land assets, housing quality or consumption that could have been affected through BRAC membership.

In the long run it can be argued that most variables are endogenous and determined by some underlying structural model. However for the purposes of this analysis we can assume that BRAC membership will not affect the explanatory variables used in the model during the four years that the organization has been in Matlab.

The 'Membership Model'

BRVO = f(ADFEM, ADMAL, AGEHHH, AGESQ, ASA, AVEDUC, BRDB, EARNR, ELECT, GRAME, HHHLBR, LANDQN, MRKTIM, SXHHH)

where,

ADFEM	number of adult females aged 15-60
ADMAL	number of adult males aged 15-60
AGEHHH	age of the household head
AGESQ	age of household head squared
ASA	one if ASA is present in village; zero if not
AVEDUC	average years of education of members in household
BRDB	one if BRDB is present in village; zero if not
BRVO	dichotomous variable; one if the household is BRAC member and zero if not
EARNR	ratio of number of earners in household to number of household members
ELECT	one if village has electricity, zero if not
GRAME	one if Grameen Bank is present in village; zero if not
HHHLBR	one if household head is a day labourer; zero if not

LANDQN	total amount of land owned (including homestead land) in decimals
MRKTIM	time in minutes to reach market place
SXHHH	sex of household head; one if male, zero if female

Land can be used as a proxy for wealth and due to its centrality in BRAC's targeting rule is an obvious determinant of membership. Land transactions are relatively infrequent and hence the assumption of exogeneity should be safe (see Pitt et al 1995 for a similar argument). The average number of years of schooling in the household was also included as better educated households are more likely to be wealthier and hence ineligible to join. Fertility decisions affecting household composition and hence also the earners' ratio can also be considered exogenous. The 'earner ratio' variable will be the outcome of two countervailing forces. Households with few earners are more likely to turn to BRAC as their source of credit; on the other hand the lack of earning members could act as a disincentive to join if there is a shortage of family labour to manage the loan investment. Moreover, the more adult females in the household the likelier it is for households to be BRAC members; this could be due to the fact that it is adult women who are targeted by BRAC (supply-side factors) and the availability of other adult females allows substitutability of a BRAC female member's traditional household tasks while she is involved in BRAC related activities (demand driven force). The number of adult males is important due to their role in managing enterprises located outside the homestead and in marketing the output.

The presence of other similar rural credit agencies in the shape of ASA, BRDB and the Grameen Bank ought to lower the probability of a household joining BRAC; on the other hand a member may try and join more than one organization if the loan size he receives from one is insufficient to meet his/her investment needs. However in practice multiple membership is rare. The distance from the market variable and electrification were included as the 'village infrastructure' variables after preliminary tests had excluded others such as irrigation and distance from roadside due to high degrees of collinearity with other variables in the model.

The dependent variable is a dichotomous 'membership' variable where a BRAC member household is given the value 'one' and all other households are 'zero'. Logistic regression is estimated in view of the nature of the dependent variable (see Maddala 1983).

This paper estimates this equation for two different 'sample groups'.

Sample 1: This sample investigates the determinants of membership using all BRAC members (both TG and NTG) and TG non members in the ten RDP villages. There are 1069 households which fall in this category.

Sample 2: This sample is even more homogenous than the first as it includes only target group BRAC members and TG non members in the ten RDP villages (see Table 1 and the discussion on the similarities and differences between the two groups). There are 908 houscholds in this category.

Results from the 'Membership' Model

Table 2: Results of Logit Estimation on BRAC's Membership

Variables	Sample: BRAC members and TG non members (n=1069) Coefficient estimates	Odds ratios	Sample 2: BRAC TG members and TG non members (n=908) Coefficient estimates	Odds ratios
Constant	0.363	–	–0.144	–
ADFEM	0.212**	1.24	0.135	1.15
ADMAL	–0.154	0.86	–0.107	0.90
AGEHHH	0.037	1.04	0.065*	1.07
AGESQ	–0.0004	0.9996	–0.008*	0.9993
ASA	–0.302**	0.74	–0.349**	0.71
AVEDUC	0.031	1.03	–0.086	0.92
BRDB	–1.414***	0.24	–1.285***	0.28
FARNR	2.611***	0.07	–2.687***	0.07
ELECT	1.119***	3.06	1.210***	3.35
GRAME	0.058	1.06	–0.239	0.79
HHHLBR	–0.167	0.85	0.002	1.02
LANDQN	0.012***	1.01	–0.005	0.99
MRKTIM	0.253	1.29	0.461	1.59
SXHHH	–0.066	0.94	–0.152	0.86

(Contd.)

(Continued)

Variables	Sample: BRAC members and TG non members (n=1069)		Sample 2: BRAC TG members and TG non members (n=908)	
	Coefficient estimates	Odds ratios	Coefficient estimates	Odds ratios
Correct prediction %	65.300		63.600	
Mcfadden's R squared	0.113		0.075	
Maximum likelihood value	1311.600		1152.700	
Likelihood value when all coefficients equal zero	1478.500		1246.100	

Notes : ***variable significant at 1%, **variable significant at 5%, *variable significant at 10%.

In the first sample those with larger land holdings and fewer number of earners to household size are more likely to be BRAC members. A one decimal rise in land leads to a one percent increase in the odds of being a member[7] and a unit rise in the earner's ratio leads to a 93% fall in the same odds. These significant coefficient estimates can be explained by the fact that NTG members in BRAC's credit group have significantly larger land holdings compared with the target group and by the fact that amongst the poor those households with fewer people earning have greater incentives to join BRAC. A greater number of adult females in the household leads to a 24% rise in the odds of being a member; the availability of substitutes for the female member's traditional homestead tasks is a plausible reason. Moreover the presence of other NGOs appears to lower the probability a household will join BRAC. However average education, age, sex and occupational status of household head cannot be used to predict membership.

In the more homogenous second sample the main result that emerges is a confirmation of the earners ratio hypothesis, namely that in a sample of eligible households in BRAC RDP villages, households with a lower ratio of earners to members are more likely (p = 0.0002) to join BRAC. The age and age squared coefficients

---

[7] A simple transformation p(1-p)b suggests that a one decimal increase in land leads to a 0.3 percent point increase in the probability (not odds) of membership, where 'p' is the average probability of membership for this sample of households evaluated at the means of data.

suggest that the probability of membership rises with age of household head but then declines beyond a certain age. The sex of household head and land owned two variables whose mean values differed significantly amongst the two groups considered in this sample emerge as insignificant at the 10% level in explaining membership in this analysis. The presence of other NGOs is similar to the first sample. In both samples the probability of a household joining BRAC rises if the 'BRAC village' has electricity. Further implications of these results will be discussed later in the paper.

The Depth of Participation

The previous section discussed the characteristics of households who join BRAC micro-credit programs vis-à-vis those who are eligible but do not. Our attention will now turn to the characteristics of the households who are BRAC members and are actively involved in its credit program. The data will be drawn from the ten BRAC villages surveyed in Matlab and the indicators of 'active' participation will be restricted to credit activities as opposed to participation in terms of a members role in other aspects of the program.[8]

The indicators that will be used to measure 'participation depth' are whether the household ever borrowed or not, the total credit obtained from BRAC for those who did borrow, the average loan size, the average number of loans taken, a measure of 'loan concentration' (i.e. total borrowed as a proportion of length of membership) and participation in BRAC's 'sector programs'[9]. Table 3 shows the extent, if any, of differentiation between eligible and non eligible BRAC members in terms of different indicators of 'membership depth'.

The comparison of means suggests that NTG members borrow significantly larger amounts compared to TG members in terms of average loans; the cumulative borrowed figure is however not significantly different. The average number of times a household has borrowed from BRAC is only just not significant at the 10% level

---

[8] This is of course a limitation of the analysis. Participation takes many forms eg in VO management committees, as small group leaders etc. However since the main thrust of BRAC's RDP is its lending operation, 'credit participation' was considered a reasonable proxy for overall participation.

[9] BRAC's sector program loans (poultry, livestock, fisheries, sericulture, social forestry and vegetable cultivation) are complemented with input supply, training and marketing support. Due to the need to go on training courses and the greater interaction with BRAC staff a member taking a sector loan could be considered a more active participant.

and interestingly we find that TG members appear to borrow more frequently. In terms of 'loan intensity', BRAC NTG members score higher but again not significantly so, compared to TG members. The sector program loan is interesting as it suggests that at the 12% level TG members are more likely to take a sector program loan.

Table 3: Differences in 'Membership Depth' According to Eligibility in the Ten BRAC Villages Surveyed in Matlab

	BRAC TG member	BRAC NTG member	'p' value of differences in means
% borrowed at least once	89.2 (n=378)	91.1 (n=180)	0.475
Cumulative amount borrowed from BRAC (taka)	7642 (n=335)	7788 (n=163)	0.732
Average loan size (taka)	2942 (n=335)	3180 (n=163)	0.002
Average number of loans taken	2.54 (n=335)	2.39 (n=163)	0.125
'Loan intensity' (cumulative loan/ membership length)	403 (n=335)	414 (n=163)	0.736
% took 'sector loan' at least once	12.4 (n=378)	8.0 (n=180)	0.120

A preliminary message that emerges from these figures is that apart from average loan size, there does not appear to be a clear-cut difference in the depth of participation between the two groups. However one needs to consider that the full extent of the potential target/non target group spread is probably not revealed in the above table due to the limit on loan sizes set by BRAC and the limited average length of membership. Disbursement ceilings are set by the number of times a member has borrowed though the activity she wishes to invest in and her previous borrowing record are also considered. Moreover, the average length of membership is just under two years and hence the full extent of the differences in 'participation depth' cannot be explained at this stage. However, multivariate analysis may shed some more light on the factors influencing the depth of participation and this is what the next section addresses.

The Determinants of 'Membership Depth'

Reduced form equations[10] were estimated for the different dimensions of 'participation depth' (see Table 4) Steps similar to the 'membership model' were taken prior to settling for a particular model, based on the 'intuitive appeal' of the explanatory variables, quality of fit and other specification tests[11]. The 'participation depth' model introduces two new variables length of membership[12] and BRAC eligibility status[13].

Depth = f(ADFEM, ADMAL, AGEHHH, AGESQ, ASA, AVEDUC, BRDB, BREL, EARNR, ELECT, GRAME, HHHLBR, LANDQN, MEMLENG, MRKTIM, SXHHH)

The variables which appear significant from this exercise are sex and occupation of household head, presence of other credit delivering agencies, electrification and length of membership. Being headed by a female appears to be a constraint to the amount a household borrows; the coefficient is both highly significant and relatively large when compared to the other variables. Households headed by manual labourers are also likely to borrow smaller amounts both as a proportion of the number of loans (average loan) and membership length ('intensity'). Moreover, manual labourer households are also less likely to take part in sector program activities; however this finding is contradicted by the highly significant BREL coefficient which suggests that eligible households are more likely to have taken at least one sector program loan This puts into doubt the impression that the initial acceptance of new technology and non-traditional enterprises is more likely to come from the 'moderate poor'

---

[10] One limitation of the analysis for the 'membership depth' model is the fact that the sample of BRAC members may be drawn from a truncated distribution; this possibility has not been taken into account in the estimation of the reduced form (see Maddala 1983).

[11] The scatterplot of standardised residuals against standardised predicted values was inspected to check for 'linearity'. Normality was tested using a normal plot (cumulative probability of standardised residuals in the case of normality against the cumulative probability of occurrence in the actual residuals). The test for heteroscedasticity was done by plotting the standardised residuals on a histogram and inspecting whether the plot approximates normality.

[12] Length of membership in months is coded MEMLENG.

[13] BRAC eligibility (BREL) can be seen as an interaction term between land owned and the manual labourer dummy since eligibility is a combination of these two factors. BREL was used and then excluded from the final 'membership model' as the collinearity between the terms made land, occupation and eligibility variables all insignificant.

Table 4: Regression Results on Selected Indicators of Participation Depth

Variables	Borrowed at least once Logit coefficient	Borrowed at least once Odds ratio	Sector program participant Logit coefficient	Sector program participant Odds ratio	Log of total borrowed	Log of average borrowed	Log of loan 'intensity'
						OSL estimates	
CONSTANT	-0.476	–	-6.071***	–	8.369***	8.091***	5.321***
ADFEM	0.356	1.43	0.015	1.01	0.074*	-0.011	0.051
ADMALE	0.406	1.50	-0.279	0.76	0.189	0.012	0.012
AGEHHH	0.049	1.05	0.088	1.09	-0.017	-0.004	-0.012
AGESQ	-0.000	0.99	-0.001	0.99	0.0001	0.0006	0.0001
ASA	-0.199	0.82	0.108	1.11	0.026	0.055*	-0.113
AVEDUC	-0.057	0.95	0.148	1.15	0.002	-0.003	0.036
BRDB	-0.310	0.73	0.617	1.85	0.122	0.001	0.234**
BREL	-0.153	0.86	1.108***	3.03	0.057	-0.017	0.079
EARNER	-0.282	0.75	-3.062	0.05	-0.211	-0.001	0.067
ELECT	0.634	1.88	-1.007*	0.37	-0.165*	-0.068	-0.227*
GRAMEEN	-1.086*	0.34	-0.418	0.66	0.072	0.083*	-0.030
HHHLBR	0.308	1.36	-0.763*	0.47	-0.113	-0.056*	-0.138*
LANDQN	0.002	1.00	0.003**	1.00	0.0004	0.0004**	0.0008
MEMLENG	0.015	1.02	0.075***	1.08	0.019***	-0.002	–
MRKTIM	0.062	1.06	-0.208	0.81	0.022	-0.047	0.058
SXHHH	-0.034	0.97	0.015	1.01	0.308***	0.032	0.448***

(Contd.)

(Continued)

	Borrowed at least once			Sector program participant		Log of total borrowed	Log of average borrowed	Log of loan 'intensity'
	Logit coefficient	Odds ratio	Logit coefficient		Odds ratio		OSL estimates	
Variables								
Adjusted R squared						0.10	0.03	0.06
% correct predictions	89.6	—	89.0					
Initial log likelihood	368.0	—	378.4					
log likelihood Maximised	334.1	—	344.9					
McFadden's R squared	0.09	—	0.09					

*** variable significant at 1% ** significant at 5% * significant at 10%.

Note : 'Loan intensity' is defined as total borrowed from BRAC divided by membership length; hence the MEMLENG variable was omitted from the model.

Micro-Credit Programs 243

households. The reason for this may be due to BRAC's intensive presence in terms of input supply, training, credit delivery and marketing of the sector program micro enterprises thus allowing the poorer households who do not have the ability to independently access the extension services or to market these products a chance to invest in these activities. Demographic variables such as the number of adult males and females in the household, the ratio of earners, age and age squared of the household does not seem to influence the various indicators of depth. The educational variable used, average years of schooling[14] in the household, also do not have much influence on the depth of participation. The village infrastructure variables and the presence of other NGOs variables present mixed results; it seems that in villages where both BRAC and Grameen operate, BRAC members have a lower probability of borrowing but those who do take larger loans, whereas ASA's presence seems to stimulate average loan sizes borrowed from BRAC and BRDB's presence does the same for 'loan intensity'.

Concluding Discussion

This section looks at the implications for micro-credit program design of the first two parts of the paper, namely the 'heterogeneity' of credit group members and the factors affecting membership and active participation in BRAC's RDP.

There are two approaches that credit agencies targeting the rural poor can take with regard to the issue of 'non target group members'. One approach is to gradually exclude existing NTG members from the group and replace them with eligible households, taking the view that the large unmet demand for credit amongst the poor ought to take priority. In line with this view steps to make targeting more rigorous can be taken so as to minimize NTG inclusion in new areas.

However the second approach is that these NTG members ought to be retained within the group for a number of reasons. The first is the focus on financial sustainability of the organization delivering credit. Even though average membership length is only twenty three months and there are strict ceilings on loan disbursement, differ-

---

[14] Regressions were run with other educational variables namely education of the household head and average education of adult females in the household. They were both insignificant in the analysis. Average education of the household as whole was chosen due to the fact that the management and marketing of loan financed investments are known to be a joint affair involving several members of the household.

ences in average credit borrowed are apparent between NTG and TG members. It is likely that over time this gap will grow and hence larger loans can be delivered and greater interest revenue per loan generated from these NTG households once loan ceilings are lifted. However the 'loan absorptive' capacity of these NTG households is still a matter to be looked into which will determine the full extent of the revenue earning potential for a micro-credit organization. Furthermore, incentives to tap the considerable savings potential (see Table 1) of the NTG group could be devised (e.g. a two tier interest rate system of current account savings schemes with free access complemented by 'deposit accounts')[15] in order to make the lending agency more sustainable.

Secondly, these larger loans can be used to create employment opportunities for the poorest who are less likely to participate actively in credit programs. However this objective may have to be an integral part of the loan sanctioning process before large loans for medium/large enterprises are agreed upon if job creation for the poorest is actually to take place. This type of 'safety net' provision would be in line with overall 'growth with equity' objectives pursued at the macro level. BRAC is in the process of initiating a new project lending to 'graduated' or NTG members (the average loan size disbursed under this new project will be at least ten times more than the current RDP average) with proven entrepreneurial ability in order to create medium/large scale enterprises prioritizing labour intensive enterprises.[16]

A further reason for retaining the NTG group is to do with a possible reason for this group being included in the first place. Pressures exerted by target group members themselves to include some influential NTG households may compel program administrators to turn a blind eye to the official targeting criterion.[17] Hence in order to enable a village organization to be socially acceptable a few NTG members may have to be included and retained. Whilst the

---

[15] See Zaman et al (1994) for a discussion on flexible savings schemes piloted by BRAC's RDP.

[16] In line with this alternative repayment incentives need to be devised in view of the greater risk associated with lending large sums. A mixture of both 'formal' and 'social' collateral based loans can be considered for non target group households. In other words small peer groups of similarly endowed households can be formed to monitor each others loans as well as using assets to secure the loan. Moreover staff supervision and monitoring of these larger loans probably need to be even stricter than at present.

[17] This is a view expressed during conversations with BRAC program administrators and field staff.

general 'community approach' has been tried and discarded by BRAC due to the elite benefiting most, the 'target group' approach may have to be flexible enough to incorporate a number of socially influential 'middle income' households, in order to maintain a link with the other socio-economic classes in the village.

Another finding which is significant for program design is the relative importance of the earners' ratio variable in determining membership. Since the number of earners is not currently part of BRAC's targeting criterion the fact that households with fewer earners are more likely to join BRAC appears to be a 'demand-side' phenomenon. Hence if this variable is added to the current land and occupation based eligibility criteria, a greater number from this particularly vulnerable group (i.e. households with few earners to members) will be included in RDP activities.

Female-headed households are another vulnerable group for whom there is evidence that barriers to RDP membership exist (Hossain and Huda 1995, Evans et al 1995). Table 1 indicated that when target group BRAC members are compared to a control group of target non members there appears to be a significant under-representation of female-headed households in BRAC's RDP. Moreover, the evidence presented in this paper suggests that female headship also curtails active participation amongst those who do join. Policies have to be designed to meet the needs of this particularly vulnerable group bearing in mind the socio-cultural constraints in involving such households in credit programs (Hossain and Huda 1995). The additional 'purdah' barriers imposed on females without a living husband which restrict both mobility and the type of loan investment have to be catered to as well as the breakdown of the traditional family based social security mechanism. Hossain and Huda in their work on female headed households in Matlab feel that '...*the social rules about what work women can do have not changed at the same rate as the deterioration of the social safety net system and the outcome is the extreme vulnerability experienced by women-headed households*' (pg. 30). The significance of the occupational and wealth variables suggest that whilst the ultra poor measured in terms of land-holding and occupation may be able to join BRAC these members do not participate as actively as other members[18].

---

[18] Whilst the BREL variable may itself not be significant the two components of BRAC eligibility ie manual labour and land size are.

If credit programs are to serve the interests of the most vulnerable in society, certain institutional features may have to be changed to accommodate this group. Initial loan repayment has to be staggered in line with the time frame of loan investment returns. Installment payments could be made monthly though the effect on overall 'credit discipline' has to be monitored.[19] Compulsory savings requirements have to be eliminated or reduced to a minimum if the poorest are not to be deterred from taking part. Easy access to savings and consumption credit[20] in order to meet emergency needs could be initiated. Moreover the 'credit-plus' approach of supporting loans with a training, technical assistance and marketing package could be more relevant to the needs of the poorest.[21] However aside from changing the design of credit programs the issue of providing employment opportunities as an alternative to credit, discussed earlier, has to be strongly emphasized.

Whilst the provision of 'flexible and diverse financial services' is in vogue in current micro finance thinking,[22] the extra administration costs of any new policies directed at a subsection of members has to be weighed up against their benefits. For instance the 'credit-plus' approach which may be more suited to the needs of the poorest involves more costs for the implementing organization than the 'minimalist approach'. On the other hand, initiating separate policies for the marginal poor in terms of larger loans and savings incentives may generate sufficient revenue for the micro finance organization to cover the additional costs over time. Hence further research is needed on this crucial issue in order to assess the 'optimal' degree of flexibility given the possible trade off between improved services for a heterogeneous client group and the financial sustainability of the micro finance institution.

---

[19] BRAC's RDP has launched an experimental monthly repayment system in selected branches to monitor the effect on loan repayments.

[20] Table 1 showed how target group households face a higher incidence of food shortages compared to non target members.

[21] BRAC's IGVGD program caters to the needs of the most destitute rural women for whom traditional credit programs are not the answer. This program works with women who are given monthly wheat relief rations, provides training in homestead poultry rearing and progressively offers concessional loans with a monthly repayment requirement. These members are gradually absorbed into the mainstream RDP program and offered larger loans. This mechanism is designed to facilitate the entry of the poorest into regular credit programs and acts as a transition from a relief to a longer term development program.

[22] See Wright (1996) for a discussion on flexible financial services in the context of Bangladeshi rural finance organizations.

Chapter 11

THOSE LEFT BEHIND: A NOTE ON TARGETING THE HARDCORE POOR

Syed M Hashemi

Introduction: Poverty in Rural Bangladesh

Even a casual observation of some of the macro indicators of rural Bangladesh reveals the intensity of the prevailing poverty. More than half the rural population have a daily caloric intake that is less than the recommended minimum, while a third receives less than the critical minimum needed to maintain body weight. Anthropometric measures reveal that three-fifths of rural children suffer from chronic malnutrition (Hossain, 1989). Unemployment and underemployment in rural Bangladesh are critically high due to high landlessness (more than half of rural households are functionally landless); low absorptive capacity of agriculture; and high population growth (a million entrants into the labor force every year). In addition the rural poor survive amidst the endemic insecurity stemming from their lack of social status. The disciplinary mechanism of the prevailing patriarchal structure ensure the acquiescence of the poor to the dictates of the dominant kin group of faction. Transgressing the boundaries of "proper conduct" can result in eviction from land, expropriation of assets, physical violence and even imprisonment.

Women are the hardest hit under poverty. Anthropometric studies show that female children in rural Bangladesh suffer from greater malnutrition than do male children. The death rate for girls between the ages of one and four is higher than for boys (even though the infant mortality rate for males is higher) suggesting clear gender discrimination immediately after birth (Jahan, 1989).

Grameen Bank and Micro-Credit: The Rationale

The landscape of poverty, powerlessness and gender subordination in rural Bangladesh forms the contextual basis for Grameen Bank and indeed the micro-credit model in Bangladesh. The famine of 1974 provided the urgency for Professor Muhammad Yunus to look for alternatives. He discovered that while the credit market was the scene of the most brutal exploitation of the poor (with high interest rates leading to persistent indebtedness leading to forced sale of assets and destitution) it was also the arena where interventions where easiest for allowing the poor to break out of their cycle of poverty. The conventional banking structure however does not provide access to the poor because the poor can provide no collateral and because the overheads required for servicing loans become too high for the small size of loans that poor people require. Governmental loan programs for rural areas in turn get monopolized by the rich and powerful. Amongst the poor, women are even more discriminated against; because patriarchal norms ensure their exclusion from de facto ownership of assets and because the work that women generally engage in (home based) are not classified as economically productive. Hence the need for targeted collateral free credit for the poor and specifically for women; hence the need for micro-credit.

Since its inception as a pilot project in 1976 and since its set up as a chartered bank in 1983, Grameen has spread its operations into 35,500 villages (more than half the villages in Bangladesh) and provided loans to 2.1 million members (more than 94% of whom are women). By the end of 1995 it had disbursed over US$ 1.6 billion in loans and has recorded a recovery rate exceeding 95%. The cumulative savings of members total US$ 125 million. The work of other NGO micro-credit programs are no less impressive. BRAC's RDP, Proshika and ASA have all extended nationally and together with Grameen provide credit to a quarter of all rural households.

This paper draws on the experience of Grameen Bank to highlight problems with micro-credit programs in general. While other credit programs in Bangladesh may have some distinctive features, the larger programs with emphasis on minimalist credit generally have some basic similarities that would lead them to similar problems of member selection faced by Grameen. These would include objectives of financial viability, high repayment rates, increased membership, as well as structural features such as the group mechanism itself.

The Credit Delivery Model

Credit programs target credit to the rural poor—those who are functionally landless (owning less than half an acre of land) or having assets amounting to less than the value of an acre of medium quality land or engaging in wage labor for one's livelihood. Since collateral is not required credit programs generally rely on the group mechanism to ensure effective repayments. The group mechanism transfers risks of non repayment from the program to the group itself. The problem of asymmetrical information (programs having limited information on borrowers) is resolved through selection of members by the group (screening out high risk borrowers), and through imposition of joint liability on the group. While individual borrowers receive loans, sanctions (in the form of suspension of new loans) are imposed on the collective group in the case of default by any individual borrower. Peer monitoring therefore reduces transactions costs and allows for successful implementation of targeted credit programs (Stiglitz 1990; Besley et al 1991; Aghion 1994; Matin 1995).

In the case of Grameen, groups of five (men and women separately) are formed by individuals themselves, selecting for those belonging to similar social and economic backgrounds (to eliminate unequal bargaining strength), from the same village and from those they have confidence in. Only one member is allowed per household. Generally a month of meetings involved in learning to sign one's name and learning about Grameen and memorizing Grameen's "sixteen decisions" regarding social conduct, takes place before the group receives formal recognition. This time allows for group members to develop close relations with each other and be certain about their decision to join. It also instills importance and confidence in members who realize that they have "earned" membership only after a long process of screening.

Loans are disbursed to members only with the approval of the group and the final approval of the Bank. The disbursement schedule is staggered; two members receive loans followed by another two members in a month and the last member (usually the group chairperson) in another month. Six to eight groups are integrated together in a Center. The center meets once a week (generally in the same premises in the village deemed as the "Center") in the presence of a Bank worker. Credit transactions as well as discussions on Grameen or different social issues take place at these meetings. Repayments are made at these weekly meetings and attendance is compulsory. Savings mobilization from group members in the form of weekly savings and a percentage of borrowed funds is also compulsory.

In the early years of BRAC, loans were made out after a compulsory period of "conscientization" of about a year and after members had saved up a certain proportion of the amount of loans they were applying for. Recently it seems that BRAC's village organizations have begun to be based on the group concept and loan disbursements are made out soon after membership is provided. The emphasis on "conscientization" seems to have been diluted.

Reaching the Poor

Most studies of the Grameen Bank (Hossain, 1988; World Bank, 1995) indicate that Grameen successfully targets the poor; most members being functionally landless with ownership of land below fifty decimals. Amongst the poor too, Grameen successfully targets women; 94% of the membership are women. Other credit programs also seem to have a membership comprised of mostly the poor (BIDS, EPAP 1993; Ruhul Amin et al, 1995). Micro studies also strongly indicate that credit programs have been generally successful in increasing the economic well-being of members and in many cases even their self confidence and assertiveness. However this does not seem to have contributed to any significant declines in the levels of poverty if one were to look at macro economic figures. In fact, different studies present conflicting pictures of the poverty situation. In the best scenario poverty seems to have declined very marginally over the last seven years (Rahman, 1996). This curious phenomena of increased NGO and micro-credit activity and persisted poverty may partially be explained by improvements in economic welfare of program members countered by declining conditions of those outside such safety nets.

In rural Bangladesh there is significant differentiation within the ranks of the poor. Roughly about half of the poor constitutes what is referred to as the hard core poor, who are forced to subsist on a daily calorie intake of less than 1740 calories (2112 being the calorie intake at the poverty level) and a per capita income that is less than three-fifths that of the poverty line (Rahman, 1995). While the poverty situation seems to have improved a little over the last seven years (Rahman, 1986) a little less than a quarter of the rural population still seems to be within the ranks of the hard core poor.

It seems that Grameen and similar credit programs have failed to target this group effectively, resulting in most of them remaining outside the micro-credit net. For the most part these people are so destitute that they consider themselves not credit-worthy. They do not feel they have enough resources to generate incomes to pay back loans. They therefore "self-select" themselves out of credit program membership.

An exercise was conducted in four villages where Grameen and BRAC were active to determine the reasons for target group households not joining credit programs. It was found that out of 498 target group households only 284 (57%) joined Grameen and BRAC as members. The major reason for not joining (49%) was because people felt they would be unable to pay back the loan money and would therefore be stuck with debt for which they would have to eventually be forced to sell off what little possessions they still had. They refused to be burdened with still another debt. A little over a quarter of the women did not join because of social and religious sanctions that dictated that joining credit programs and leaving the home for meetings with outside males would be a violation of social norms. Only 13% of the women said they actually wanted to join but were not accepted because other program members felt they were high risks (their husbands were gamblers and would waste the money; they would migrate out of the village; they were not good money managers; they did not get along with others). Surprisingly 19 women in Grameen villages said that the rules were too complicated and they couldn't memorize the sixteen decisions (a prerequisite for membership).

Actually what this is indicative of is not so much that Grameen and BRAC are unable to bring all poor women into their fold but that micro credit is not necessarily the way out for all the poor. Successful micro credit operations are strongly dependent on strict screening to ensure that money that is borrowed can be repaid. Groups themselves or group leaders and NGO staff are extremely careful to screen out potential risks. Households having some assets, some steady incomes (the better off among the poor) are encouraged to join. It is felt that even if program funded enterprises do not immediately generate profits or if there are some losses these households would be able to make up for it through their other incomes or through sale of assets. It is felt that destitute households would either consume all incomes from funded enterprises and thereby be unable to make repayments or would be too poor to sustain even minor losses. Poor recoveries would reflect on overall group performance and performance of NGO staff. Neither group leaders nor NGO staff are willing to take that risk. Thus, in order to ensure increasing disbursements and high repayments (the targets that are set up by programs) NGO field staff and group leaders are, therefore, almost led to screen the destitute out.

Table 1: Target Group Members not Joining Grameen and BRAC

	Rangpur Grameen	Villages BRAC	Faridpur Grameen	Villages BRAC
Total # of households	246	110	189	191
# of target group households	173	74	140	111
Program members	99	59	84	42
Target group hh not members	74	15	46	69
Reasons for not joining				
can't repay loan money	41	5	14	40
social/religious sanctions	21	7	14	16
can't memorize 16 decisions	9		10	
other members do not accept	3	3	8	3

Even though the weekly repayments schedule is set up to ensure that it is easy for poor people to make small repayments over a long period of time rather than making repayments all at one time, for the really poor the weekly repayments are also often difficult. For the destitute and for others with difficulty in making good use of the loan (investing or purchasing of assets rather than meeting immediate consumption needs), credit programs therefore may not be the answer. Other targeted programs would be required to address their specific needs.

This failure of all sections of the poor to appropriately use credit has specific implications regarding coverage of credit programs. While village after village may be covered by credit programs (extensive coverage) there would be great difficulty in ensuring intensive coverage of each village. There would be great gaps therefore in the poverty alleviating net which could effectively hold back community level gains even while individual members may be improving their economic position.

There seems to be two prevailing strategies in addressing this issue of bringing the poor in. Firstly, steps are being taken to increase local level economic activity. Programs for increasing agricultural productivity, setting up of rural industries, marketing facilities, all are geared towards increasing employment and incomes at the local level. This would then contribute to making non-program members more comfortable in seeking loans to set up new enterprises—since the increase in local demand would expand the potential for running profitable ventures. Grameen's initiatives in textiles (Grameen check and the setting up of a textile factory), in agriculture and fishery (the Krishi and Motso Foundations), in solar and wind energy (Grameen Shakti), and in telecommunications, are all examples of strategies for increasing economic activity and increasing aggregate demand, that can be facilitated through the use of micro-credit.

The other strategy seems to be in directly setting up programs specifically for the destitute.

The BRAC IGVGD Program

The BRAC IGVGD program would be an example of this. The IGVGD program is targeted to provide food and development services to the poorest women in rural Bangladesh. It attempts to improve their economic and social conditions so that they may graduate beyond their existing conditions and be able to sustain themselves above the hardcore poverty level.

The selection criteria for the VGD women specify that women from landless households owning less than 50 decimals (0.5 acre) of land, women with irregular (less than *taka* 300 per month) or no household income, women who are daily or casual laborers and women from households lacking ownership of productive assets, be selected. Additionally, preference is given to female headed households (widowed, divorced, separated, deserted women or women with disabled husbands).

Women participating in the program receive a monthly ration of wheat or rice for a period of two years. In addition, through the formation of groups women are made to save, and receive training on income generating activities (poultry rearing, sericulture, livestock raising). Credit is provided to set up these activities. It is expected that such support in income generating activity, while relief provisions are made available, will provide a longer time frame for these women to slowly gain the confidence not only to earn incomes but also to access available resources. The IGVGD program therefore combines relief handouts with training to slowly graduate women to a stage where they can be confident in becoming NGO members.

Detailed surveys of the IGVGD program (Hashemi, 1994, 1995) indicate that the VGD Program has been successful in targeting the most destitute rural women, women who generally are either excluded from NGO programs or who self-select themselves out because they are too destitute to be considered credit-worthy. The survey results also show that the wheat and savings intervention (as was previously being done by the government) alone does not improve the economic condition of poor women. The VGD Program has a positive effect on women's economic status in terms of generating higher individual income and ownership of assets as well as on food consumption and investment of the households as a whole, only if support services are provided to the women such as through the IGVGD and other NGO interventions. These interventions are also more likely to increase the social awareness of the women and their participation in public space.

Conclusion

Micro-credit in Bangladesh, through Grameen Bank, BRAC, Proshika, ASA, and other governmental and non-governmental agencies has succeeded in reaching a quarter of all poor rural households. This has had significant impacts in increasing economic welfare as well as initiating major social changes in rural Bangladesh. However poverty still persists. One major reason for this may be the limits to micro-credit in effectively targeting all of the poor; more specifically in leaving out large sections of the hard core poor, the distressed. One element of the response has been to attempt to increase productivity and employment and through this to increase the credit absorptive capacity of the local economy. The other part of the response, strategies for better targeting of the destitute, now requires greater emphasis. BRAC's IGVGD program seems to be one such experiment. Grameen's attempts at bringing destitute defaulters back into the credit fold in Rangpur through "share-raising" of goats is another such experiment. However more creative thinking is required for this to be totally addressed. As a first step it is imperative that those who are being left behind in prevailing micro-credit strategies receive top priority in our program agenda, in our conceptual framework and in our analysis.

PART IV

MICRO-CREDIT-LIMITATIONS OF SCALE

Chapter 12

THE RENEGOTIATION OF JOINT LIABILITY: NOTES FROM MADHUPUR[*]

Imran Matin

"The practice of denying credit to all group members in case of default is the most effective and least costly way of enforcing repayment" (The Role of Groups and Credit Cooperatives in Rural Lending by Huppi, Monika and Greshon Feder in World Bank Research Observer, 5 (2):187).

"They (referring to a group credit lending institution) do not need police to compel repayment: we (centre members) have been doing it so long... it can't be done any longer" (A group credit member in Madhupur).

The group credit arrangement practised by the "Of non-poor-For poor" (Copestake, J., 1994) credit institutions[1] has drawn enthusiastic response from many. Very broadly, we observe three major streams in this enthusiasm:

School	What Attracts Them
Ohio School	a. Emphasis on subsidy-free interest
	b. Emphasis on financial sustainability
Imperfect Information	a. Contract Design Features that smoothen imperfect information constraints
Poverty	a. Outreach efficiency
	b. Poverty (alleviating) impacts

---

[1] Though usually referred to as 'financial' institutions and at least one as a 'bank', I prefer to describe these as "credit" institutions as they are principally lenders

In this paper, I concentrate on exploring the enthusiasm of the imperfect information school around group credit arrangements. In the first section, I discuss the existing theoretical understanding of the mechanics of group credit arrangements and how it is supposed to smooth imperfect information problems. Subsequently, I deal with the problems with the existing theoretical models indicating their biases and finally I try to show the ruptures of the system.

The Pristine Story

The imperfect information school maintains that because credit transaction is not a spot transaction, it suffers from acute imperfect information problems. They come in three forms: hidden information, hidden action and enforcement constraints. These constraints are however asymmetrical in their severity, implying that they are relatively more severe for a distant, non-local lender (or his[2] agents) compared to the locals who are assumed to possess a *positional information advantage* (about other locals) over the non-local lender. Thus, what is required are innovations in contract design that would give the locals an incentive to use their positional information advantage in smoothing imperfect information constraints.

Joint liability, as popularly understood, is a contract in which the provision of the private good (e.g.: an individual's access to credit) is made conditional on the provision of the public good (group repayment). This is seen as an effective and least costly incentive making the borrowers use their knowledge about each other in screening the 'right' people (thereby smoothing the hidden information problem), engaging in peer monitoring (thereby reducing the hidden action problem) and exerting peer pressure (thereby alleviating the imperfect enforcement constraint).

In what follows, I summarize two important theoretical attempts that use the above logic in modeling group credit arrangements:

In one of the very first theoretical attempts in developing a 'general' theory of peer monitoring (Stiglitz, J.E, 1990), Stiglitz shows that *because* joint liability results in high repayment, the lenders will be ready to offer a lower interest rate to the participants in such an arrangement, than otherwise. The important point that

---

[2] Throughout the paper I use 'he' for the lender and its agents and 'she' for borrowers for obvious reasons.

Stiglitz makes is that the resulting interest rate will be lower than the reservation interest rate of the participating borrowers. This leads to welfare gains for the participants.

Another very important (but often ignored) condition Stiglitz attaches to his proposition is that such welfare gains would result only at low levels of co-signer liability which Stiglitz assumes will be the case due to effective peer monitoring.

It is not difficult to see the problems with Stiglitz's assumptions:

- Peer monitoring is assumed not only to be cost-free (and unproblematic) but also effective in maintaining low levels of co-signer liability. The questions are:
    - Is peer monitoring cost-free ?
    - How is this 'low' level of co-signer liability maintained? Is the process unproblematic?
- The process of enforcement of joint liability is seen as unproblematic.

The paper by Besley and Coates (1991) addresses one of the questions above: regarding the ways in which low levels of co-signer liability could be maintained. In a two person setting, they argue that joint liability (again assumed to be negotiation proof) gives rise to positive and negative externalities: the positive externality (see fig. 1, note 'a') arising from peer support, a form of informal in-group insurance in which the 'unfortunate' (defined as the group member whose return on the loan funded project is so low that repayment is not individually rational) is 'helped' by the 'super successful' (defined as the peer whose return on the loan funded project is so high that repayment for both is individually rational for her. This is because the benefits of refinance, which can be accessed only if group repayment is materialized, is assumed to be increasing in borrower's project yield). This is the well known 'privileged' group setting.

The negative externality (see fig. 1, note 'b') arises when the 'unfortunate' cannot be bailed out because the peer's project yield is in the 'intermediate' range, not high enough to 'help' (i.e. repay for both) but high enough to repay the individual due. In such an 'intermediate' group setting, joint liability leads to non-repayment.

Besley and Coates point at an interesting solution to this: the positional information advantage possessed by the peer members should enable them to develop a penalty mechanism backed by social sanctions. To understand how this is to solve the negative externality problem, let us think of a 'break-even' project yield where the borrower is indifferent between repayment and non-repayment.

Under the 'social sanction regime' (see fig 2), a 'fairness' project yield exists (needless to say that such a 'fairness' project yield (say A^f in fig. 2) will be lower than the 'break-even project yield' without social penalty (A in fig 1 and 2)) beyond which repayment will be enforced. Below this 'fairness' project yield, there could exist group repayment solution where each borrower repays a sum equal to the benefit each attaches on future access to credit. It is clear that group repayment may be made even when a borrower's project yield is below the 'fairness' level and the peer's in the intermediate range if the sum of their contributions (which will be equal to their individual valuation of future access to credit. As the valuation function is assumed to be increasing in project yield, the borrower in the intermediate range will pay more than her due while the borrower whose project yield is below the 'fairness' level will pay part of her due) is at least equal to the group due.

All this seems very neat. But the underlying logic based on loan funded project yield assumes firstly that credit is used to finance projects the returns of which drive individual repayment decisions and secondly that this can be observed by the peers without cost. If repayment decisions are grounded on the more complex dynamics of the household economy like the saving potential of the household then the neat implications of the model becomes fuzzy, to say the least.

The Fuzzy Story

As a matter of fact, the emphasis on peer monitoring in all these models suffers from a 'project bias'. It is curious to note that the arrangement of the weekly repayment schedule is not addressed in any of these models precisely because it is not thought to make a difference: if credit is used to finance projects then the essence of the repayment game would not change because of the repayment schedule. In that case, how would one explain the high repayment rate with interest rate around 25% a year and loan capital having to be repaid within a year? If the 'project based' argument is credible then a new project would have to produce a rate of return of 125%

just to service the loan. If working capital is needed and if the family is to be fed, the required rates of return become astonishingly high (ADB, 1994: 67).

I find the concept of advances against future savings a much more realistic lens in understanding the repayment game where the credit taken is a lump sum advance (expensive, though) against the ability of the household to save in small quantities which principally make up the '*kisti*' (repayment). The high repayment of these new-found credit arrangements then is more due to the weekly repayment schedule which allows the borrowers to make use of their small savings with the centre and group structures acting as 'disciplinary mechanisms' required to 'force' the household to save in a sustained way. This in turn implies that crisis and unforeseen uncertainties which impede this 'sustained savings' process is also likely to affect repayment.

Then, what about peer monitoring? If repayment decisions are based in the dynamics of the household economy rather than cost-free observable project yields, then what do the peers monitor? More fundamentally, what triggers the whole game?

We know of the interest of the lenders in joint liability. It compels repayment and reduces their transaction cost. But, joint liability is hardly something that the borrowers gain from. What they agree to contractually is the *ex-ante* tying of the private good with the public good. As long as the lenders can enforce this tying, the borrowers would choose the (private) least costly strategy to ensure the provision of the public good.

Peer monitoring, not in the sense described in theoretical models where peer group members monitor each others' projects and design elaborate conjectures through which the problems of the repayment game is solved, but where peer group members monitor each others' loan repayment profile based on which they internally ration credit, is very costly.

Screening is hardly effective firstly because it is a result of collective action and bargaining the result of which is not readily predictable or necessarily desirable. Secondly, the space in which screening takes place is not 'neutral' and thirdly, because screening does not preclude strategic behaviour in the future.

Peer support rarely occurs and even when it does, it covers a few partial shortfalls (many of which are actually engineered by the bank worker driven by the '*kisti* stress'. A very popular strategy is to use members' weekly savings or group fund loan instalments to cover such manageable short-falls). In many cases, the borrower facing instalment shortage would take a short period (very), instant, high interest bearing, informal loan from non-borrowers (in the fear that if any centre member knew, she might inform the bank worker which might affect her future loan size) or sometimes from very close borrowers[3] (whom she can trust), if possible.

What does work (but see below), however, is peer pressure and contrary to Besley and Coate's simple world of 'project yield fairness', such pressures do not take into account the circumstances of non-repayment (unless it is due to socially significant events like death or very serious illness. Death actually leads to quite quick settling of debt to a large extent for cultural reasons). Any non-repayment of an instalment is made to increase the transaction cost for all borrowers in the centre (not only the group) as the bank worker quite rightly feels that the centre would be a stronger and more effective vehicle than the group to pressurize. The increase in the transaction cost takes various forms: making all the borrowers wait until the whole centre's repayment is made, delaying sanctioning of loan proposal unless centre repayment record is perfect or unless the centre takes satisfactory steps to 'solve' the non-repayment problem etc.

Two important points are to be noted from above: firstly, it is staff pressure that triggers peer pressure. The question then is, to what extent is it sustainable? Secondly, joint liability as usually understood, *viz.* termination threat where all the group members are denied access to future credit if full group repayment is not made, is not totally correct. Let me elaborate on the second point first:

When repayments are made on a weekly basis, it is obvious that the termination threat understanding of joint liability is not appropriate. Under such a regime, group (actually the centre) failure of complete weekly repayment cannot be sanctioned by a termination threat. As pointed out above, it is sanctioned by staff-induced peer pressure in case of '*kisti* problem'. This implies that the observation that termination threat is not enforced does not necessarily imply that joint liability is not operational (Jain, P, 1996 makes this mistake). Now, I turn to the first question raised above.

---

[3] The reported contract is loan of 100 Tk. for 3 days with an interest of 25-30 Tk and loan of 100 Tk. for 15 days bearing an interest of 50 Tk.

The Ruptures and Renegotiation of Joint Liability

To start with some descriptive figures: for all the centres in four villages[4] of Grameen Bank in Madhupur, Tangail, operating since 1980, I calculated the following as of July 1996[5]:

Centre	Total Drop out (from official records)	No. of inactive borrowers 1[6]	No. of inactive borrowers 2[7]	No. of regular borrowers	No. of overdue loans in 1995[8]	Repayment Rate of the overdue loans as of 31.12.95[9]
A	5/32	8/32	16/32	16/32	10/37	53.4%
B	4/35	10/35	31/35	4/35	25/36	46.6%
C	6/20	4/20	7/20	13/20	8/26	60.51%
D	1/39	9/39	25/39	9/39	22/43	28.55%
F	0/25	6/25	19/25	3/25	17/22	34.22%
G	12/40	4/40	11/40	20/40	32/44	32.63%
H	7/40	2/40	27/40	7/40	20/55	34.95%
I	4/40	1/40	4/40	30/40	33/54	37.52%
J	6/25	0/25	15/25	10/25	19/35	55.86%
K	5/30	2/30	7/30	6/30	28/41	42.10%
L	6/30	4/30	12/30	13/30	24/35	50.21%
Ave.	56/329 (17.02%)	50/329 (15.19%)	174/329 (52.88%)	131/329 (39.81%)	238/428 (55.60%)	

In all the centres in these four villages the repayment rate has fallen drastically and the numbers of inactive borrowers have risen. The recorded drop out figures, however, fail to capture this trend. Many overdue borrowers have overdue amounts much larger than their individual contributions in the Group Fund. Many of the borrowers who have overdue loans have stopped repaying altogether, some repaying part of their dues as and when convenient and a few

---

[4] Jota Bari, Bipro Bari, Teki and Pirojpur.

[5] As I have not yet obtained the July 1996 statement, I do not have latest figs. for col 5 and 6.

[6] Borrowers who have not taken out any loan in 1995 in spite of not having any past bank dues.

[7] Borrowers who have not come to centre meetings for a year or more obtained from survey questionnaire

[8] No. of loans taken in 1994 and not repaid by 31.12.95. Obtained from yearly statement of 31.12.95.

[9] Repayment rate of loans reported in col.6. Calculated from yearly statement of 31.12.95.

remain to be good repayers. Those who are still good repayers are getting new loans. Staff pressure and the concomitant peer pressure is almost non-existent. The whole system is now operating on the basis of individual liability.

Why and how has joint liability faltered? An interesting feature of this system is that when the 'unzipping' process begins, it leads to a snow-ball effect. This means that the distinguishing variables between the groups become fuzzy. Thus, my analysis is based more on the processes and the dynamics of the rupture based on detailed discussion with centre members.

The argument that joint liability is merely a theoretical construct and never operated in practice is difficult to sustain. True, joint liability in the form of termination threat hardly was carried out but there is ample evidence to show that joint liability in the form of staff pressure induced peer pressure had been in place until recently in the study area.

A potential trigger factor in weakening joint liability seems to be the centre's inability to enforce internal credit rationing. One would expect that joint liability would serve as an incentive for the borrowers to use their information advantage for internal credit rationing. Despite the centre members having the necessary information to do this, revealing this information is considered problematic, especially when the loan size is large. Moreover, the centre is not a 'neutral' space.

Thus, the only way in which joint liability works is via staff pressure induced peer pressure which is directed at the centre to solve '*kisti* problem'. This works as long as the '*kisti* problem' is manageable, that is when there are a few with '*kisti* problem' and when the '*kisti*' is small. But this is not sustainable.

As pointed out above, the channel through which staff pressure triggers peer pressure is by increasing the transaction cost of on-time repayers of the centre. As loan size increases, more inefficient loan sizes are disbursed which increases the cost of peer pressure by increasing the transaction cost (as more and more borrowers become irregular) for on-time borrowers and decreasing the effectiveness of peer pressure (as more and more borrowers need to be pressurized). A point could be reached where eliminating 'problem borrowers', which is often used to 'clean' the centre becomes difficult as a borrower's overdue amount far exceeds her group fund contributions. As numbers of on-time repayers decrease, an 'unzipping' effect is likely. This would render staff pressure induced peer pressure increasingly ineffective.

Zeller, M and M. Sharma, 1996 also find loan amount to be significantly positively associated with default while the square of loan amount, a variable trying to capture the effect of increased cost of default (due to incremental penalty rate of interest) when loan size is large, is found to be insignificant. They also find the variable ration (probably external rationing by the lender) to be significantly negatively associated with default. Interestingly, too strict rationing, captured by the variable squaring ration is significantly positively associated with default. That loan size determination in a multi-period setting could be a cause of failure of joint liability has also been pointed out by Aghion, B.A, 1994.

Those Who Get it Wrong

Even if the above dynamics are counteracted by say continuous vigilant 'cleansing' of the centre (many centres of Buro Tangail, ASA and SSS have been rigorously 'cleaned' and some have been 'vanished'), this process would tend to sift out the poorer customers, more specifically those who have been unable to increase their savings capacity to absorb increased debt. The more worrying trend is that of taking on increased credit (which can be easily obtained if repayment is made) beyond savings capacity. This trend could be triggered by a combination of repayment difficulties (the schedule of which is negotiation-proof), failure to 'craft' the loan into something useful, crisis, peer pressure and increased informal debt burden. In the end, these drop-outs (sometimes they cannot choose to drop out as their debt burden cannot be covered by their accumulated forced savings with the lender) often emerge with few or no financial assets. I have come across cases where such a downturn leads to more expensive access to informal credit sources as their credibility is impaired.

I do not see any way in which joint liability group credit contract can induce peer members to use their positional information advantage for internal credit rationing. Any suggestion for greater external monitoring and rationing would have to show that the external lender has ways of accessing and assessing the information required to enable it to do so in a cost effective manner. Otherwise the results could be counter productive.

Borrowing lump sum sums against a stream of small savings under an arrangement that provides the discipline required for the purpose has been useful for millions. But the system shows ruptures because of its assumption that the continuous injection of increased credit leads to an ever expanding virtuous cycle of increased savings capacity. For many, such a 'promotional' view of credit is not appropriate. It is these people who either do not get access or who are spat out by the system over time. A reconsideration of existing models and conceptions is thus required if an extension of formal financial frontiers is to be achieved.

Chapter 13

POVERTY, PROFITABILITY OF MICRO ENTERPRISES AND THE ROLE OF NGO CREDIT

Rushidan Islam Rahman

Introduction

Micro enterprises have existed in this country throughout history. Many of these were caste-based and survived through generations, without much change in technology, skill or scale of operation. Since the objective of such activities was survival and generation of employment for the family labour force, these were not viewed as entrepreneurial activities. The activities were taken up as a routine matter without having to undertake a conscious effort or much risk-taking. Therefore the terms 'family activity' or 'self employment' were adequate to describe these enterprises.

The recent enthusiasm about micro enterprises has been the outcome of concern related to the problems of poverty and under-employment/unemployment and the belief that the generation of non-farm self employment can solve the problems of both employment and poverty. In densely populated developing countries like Bangladesh, the large scale manufacturing sector has been unable to absorb the growing labour force. Low and declining land-man ratio displaces rural workers from crop production. Development theories postulated that industrialization would progress through the use of cheap labour attracted from the rural areas (Lewis 1954, Ranis and Fei 1964). This did not happen in reality and large scale industries proved to be inadequate even for absorbing the growing urban labour force, not to speak of drawing the surplus labour from the rural areas. Whatever be the reasons behind such performance of the economy, this scenario implies a crucial role for micro enterprises (ME) in the labour absorption process.

A conscious effort to develop non-farm activity led to the use of the term 'micro enterprise'. Here the core element is entrepreneurial initiatives, which meet with a small initial investment to utilize family labour force. Such enterprises may gradually raise the size of investment and draw in hired labour. The essential element of being a micro enterprise will continue as long as the main objective of the enterprise is to utilize family owned productive resources.

Poor households possess little land or financial capital. Micro enterprises may therefore depend on finance from outside sources. The land frontier has already been reached in this country and the possibility of land redistribution is a sensitive issue. Financial capital is expected to provide an alternative form of productive resource to which the poor may get access through credit. Optimism about the role of credit in breaking the vicious circle of 'low income, low saving, low investment, low income' has been expressed by the NGOs of Bangladesh involved in micro-credit operations.

In this country, NGO activities during the eighties have taken up the dimension of an independent movement of poverty alleviation. Generation of self employment through provision of credit has been the major strategy in this movement. During the initial years NGOs have been lauded for their success in this endeavour.

As the NGOs have entered a phase of expansion, they face new challenges. Their activities and strategies are receiving more critical evaluation during recent years. Allegations on the departure of the NGOs from their own set of rules and the leakage of funds to non-target group members have been voiced from the early days of NGO activities (Kamal 1996, Hossain 1988, 1985; Rahman 1991). The other major concern about micro-credit is related to the outcome of the process, that is whether the poorest households have benefited from NGO credit (Rutherford 1995, Rahman 1996). Since poverty alleviation has been the major objective of NGO credit, concerns related to their ability to reach the poor need to be examined.

A large number of studies have assessed the role of NGO credit in poverty alleviation. Most studies demonstrate a positive effect of micro-credit in this respect. Whereas the availability of NGO credit enhances income among the poor and with continuous membership of NGOs, some of them have moved out of poverty, there is still a group of 'extreme poor' households who have not been reached by the NGOs. Even if an NGO branch operates in a village for a few years, not all the extreme poor households in that village get access to NGO membership. Recent concerns over such exclusion of the poorest households is in line with the concern about the heterogeneity among poor. Those living below the 'poverty line' are located at various distances from it. It may be easier to lift those who are closer to the line. Thus a three fold categorization of poor is possible: the extreme poor (or the poorest), the moderate poor and the borderline poor (who may be termed as 'vulnerable non-poor'). NGO success is alleged to be rather limited for the group at the bottom.

The objective of the present paper is to review the possible reasons behind the failure of NGOs to reach the poorest households. In Section 2 it will be argued that recent papers which raise this allegation have inadequately formulated the reasons behind such failure (Kamal 1996, Rutherford 1995). While the suppliers determine the access to loan for various types of clients, the demand side may play an equally important role. The scope for profitable use of micro-credit may vary depending on the initial situation of poverty and will determine the demand for credit. The profitability of micro enterprises provides a crucial link between poverty and the extent of access to NGO credit. Section 3 elaborates this hypothesis and presents case studies to support the arguments. Section 4 revisits the supply side factors behind the exclusion of the poorest households from the facilities of NGO credit. Section 5 presents some concluding observations.

As has been mentioned above, this paper is based on case study materials collected from a number of micro enterprises and does not use quantitative analysis. This has been done to get an insight into the operation of demand and supply side factors separately, which in quantitative exercises get merged into the estimation of reduced form equations. Detailed case studies for 20 enterprises have been conducted. The families involved live in a village covered by the activities of two large NGOs. While the case studies can provide richer insights into the operation of ME, and the process of use of NGO credit etc., such case studies cannot ensure statistical validity of the conclusions, which has not been the objective of the present study.

Why the Poorest are not Covered by NGO Credit: Existing Hypotheses

The criticisms against the NGOs for not reaching poor households or for their failure in strict adherence to the target group criterion have recently taken a more specific form. The exclusion of the poorest has been identified as the crux of the problem. The inclusion of marginally better off households would not be considered as a serious deviation from the goal of poverty alleviation if this was done after the entire bottom segment of poorest households had been covered with loans. Concern related to the exclusion of the poorest is quite widespread among researchers and policy makers. Yet no concrete data is available to substantiate this concern. A generally accepted notion is that the bottom five to ten per cent is excluded. Empirical data on this issue is a research priority.

There has been an inadequate understanding of the reasons behind the exclusion of the extreme poor. The most prevalent current hypothesis runs in terms of poor loan repayment prospects by the poorest households. Perfect recovery of loan is the goal of most NGOs. They consider loans to the poorest group as an obstacle to that goal.

The existing empirical evidence does not provide support for this hypothesis. Studies on the default of agricultural credit emphasise large scale default by non-poor farm households. Even among NGO clients, the 'not-so-poor' do not always demonstrate their superiority to the 'poorest of the poor' in terms of good repayment and stable group behaviours. Rutherford (1995) describes the case study of a branch of a large credit NGO 'where things go wrong'. In this branch the members default and move out of the groups because of non-congenial relationships with one another, even the borderline poor or non-poor. On the contrary it can be argued that all the extreme poor cannot be accused of loan default. Currently the large NGOs like ASA which show 99 per cent loan recovery, have a substantial percentage of extreme poor as their members. If they were not making their repayments, how could the current loan recovery figure be achieved?

Thus 'poverty' alone is not responsible for the inability to repay. There are a number of characteristics of the poor which differentiate their ability to repay. One should turn to the analysis of these factors to understand why some of the poorest households may be creditworthy while the others are not.

A special group among the extreme poor identified as suspect are the work-averse beggars, disabled, vagabonds, singers, performers etc. This group is unable or unwilling to perform hard physical labour and their exclusion is easily justified. In addition to this work-averse group, there may be households who constitute a type of "floating population". Extreme poor households who do not even possess homestead land, may be residing in others' houses where they are attached as laborers. They may easily shift their residence and move to other areas. This type of household is considered a bad credit risk and would be excluded. However, the size of the excluded group of extreme poor is much larger than implied by the above description.

Demand for Micro-Credit and Exclusion of the Poorest

A large percentage of the poorest households who remain outside the NGO credit network may have stayed away because of their own reservations and fear, which we may call "demand side consideration". The exclusion of the poorest households can be explained in terms of the interaction of demand and supply factors. To arrive at such an explanation, we now proceed to bring in the demand side. Poor households' demand for micro-credit will be generated from their ability and willingness to make a profitable investment of the fund. Therefore the question now boils down to whether for the poorest households such ability and willingness is lower than for the less poor and why this is so. To obtain an answer to this question, firstly one needs to understand the operational basis of micro enterprises. Secondly, factors associated with poverty need to be identified. Then one can see how the same factors work as constraints to the expansion of ME and account for low profitability of ME among the poor.

A detailed analysis of the determinants of poverty is beyond the scope of the present discussion. A number of studies have shown that a family's resource endowment is the most important determinant of poverty. Land, working capital for non-farm activities, non-land fixed assets, family labour and skill level of family workers are the three most important resources in this context. In addition, the dependency burden in a family determines the total consumption requirement of a household, which determines poverty (Hossain 1992, 1996). Gender composition of labour force and the age of the women are also relevant.

Operation of ME is not based on a commercial principle. Micro enterprises do not operate in a competitive market for all the relevant inputs. If they did, and capital had been the only binding constraint, then all types of households (poor and non poor) would be able to take credit and make an equally profitable investment.

Almost by definition MEs are family enterprises. This implies that the family's own resources are primarily utilized in these enterprises and they may be supplemented by hired inputs as required. As has been mentioned at the beginning of the paper, utilization of family workers has been historically the major aim of such self employment projects. This 'utilization' approach relies on the notion of easy availability of labour. To get a more accurate picture of the operation of present day MEs, the above notion should be generalized to take into account the availability of each of the inputs used. It is a rather limited view to consider the utilization of family labour input as the only objective of MEs without considering the complementarities of input use in ME. A proper view is to take into account all the major types of inputs and to identify what actually is a binding constraint for ME development.

In the following discussion we take into account the implications of the possession of family labour endowment and financial capital on the profitability of MEs. In addition, the role of land in the ME establishment process is considered.

Agricultural land is not directly related to ME development. The homestead area and the house itself are important inputs for ME. For many types of activity, the homestead area provides an important input and the size of the house and homestead land becomes a binding constraint for the expansion of a scale of activity. For example, the number of livestock that can be maintained will depend on the space available. The rickshaw puller needs a safe place to keep his rickshaw at night. Paddy processing requires a place for drying the paddy. To illustrate the importance of the homestead area, an example is presented in Case Study 1.

Though the role of family labour in ME development had received emphasis, its various dimensions need to be focused. The physical ability of poor workers and their skill may vary. On these counts there can be important differentiation among the poor. The gender composition of household's earning members may also be relevant for ME development. An important determinant of the extent of poverty of households is the size, skill and gender composition of family labour force. It will be seen that these factors also play an important role in determining the profitability of ME.

Skill and physical ability enable workers to pursue activities of higher return. For example, a person with tailoring skill or the physical ability to ply a rickshaw can generate a high rate of return. If such skilled persons are provided with loans from NGOs which enable them to own their capital equipment, the rate of return can be even higher. With such a high return, the loan from NGOs may be easily repaid. Households with such skilled labour do not usually constitute the extreme poor and they are always attractive as clients of NGOs as against households with a smaller family labour force and a larger dependency burden.

Table A1 presents case study findings on the returns to labour from eight types of activities in a village. Wide variations in the rate of return among activities are noticeable.

Table A1: Rate of Return per Hour of Labour Input in Different Types of Activities

Activity		Rate of return (Taka) per hour
Rickshaw Plying	Case 1	5.4
	Case 2	5.3
	Case 3	1.4
Oilseed processing	Case 1	5.3
	Case 2	3.5
	Case 3	2.5
Food processing	Case 1	2.5
	Case 2	2.9
	Case 3	2.8
Bamboo craft	Case 1	2.9
	Case 2	3.2
	Case 3	3.5
Paddy husking	Case 1	2.9
	Case 2	2.8
Tailoring	Case 1	6.0
	Case 2	6.2
Livestock	Case 1	0.9
	Case 2	0.8
Peddling	Case 1	8.7
	Case 2	1.8

Source : Case studies, 1996.

The extent of poverty is negatively associated with the number of workers in a family. Those with two or three working members can usually make a good balance of both ends, whereas, a one worker family with three to five dependants is usually plunged into extreme poverty. The larger family labour force makes a household suitable for NGO credit. More than one family worker can ensure a higher amount of total labour input in the enterprise and thus a higher return to capital is generated. In such multi-worker families, even if only one member is engaged in the credit financed activity, a part of the loan can be repaid from the earnings of other members. Case Study 2 and Case Study 3 present stories to illustrate these factors.

A similar argument holds for the gender composition of family labour force. Families with only female workers face a disadvantage in this patriarchal society, and male workers have an advantage in some jobs. In this respect the worst position is likely to be occupied by families with only one adult woman. However, enterprising women who run their own business without help from any male family members are also substantial in number and have been growing with support from NGOs. Stories of women entrepreneurs have received wide publicity and there is no need to repeat them here. Nevertheless, an example is presented in Case Study 4 which shows that the lack of male workers is no constraint. In fact it was difficult to find a case where a woman faced obstacles just because she did not have a male worker to help her out.

NGO credit does not require equity capital on the part of the borrower-investor. Thus the penniless can also use NGO credit and start a ME venture. Though one's own capital is not essential to start a business, it may determine the returns to labour by raising the capital/labour ratio and also by affecting the choice between fixed capital and running capital, which, in turn, determines the type of activity pursued. The logic behind this proposition needs some elaboration. Those who do not possess their own capital are likely to use a smaller proportion of the borrowed fund as fixed capital. This constraint arises when the repayment period is strictly set at one year. With such terms of repayment, if the borrowed fund is invested in fixed capital, a sufficiently high rate of return needs to be generated so that the entire amount of fixed capital can be saved and retained at the end of the year to allow the loan to be repaid out of the profit earned during the year. The effective rate of interest for loan from most NGOs ranges from 20 to 25 per cent. Therefore, to pay back the entire capital along with interest, after retaining the fixed capital, a rate of return of more than 125 per cent is required.

This implies that the poorest households are more likely to invest in activities which require little fixed capital. This limits their choice to certain processing activities and petty trading or peddling. Such choices may generate a low return to labour. Moreover, after a certain level of working capital/labour ratio is attained in these activities, the marginal productivity of capital declines sharply. ME with such a low return may not be a better choice than wage employment. Thus the poorest households, who do not have any saving to supplement the borrowed fund will be discouraged from entering into ME. Table A2 presents data from a few case studies which show that the total investment in ME, and in particular the amount of fixed capital, is lower for moderate poor households, compared to borderline poor, and is lowest of all for extreme poor households.

Table A2: Use of Capital in Micro Enterprises by Three Categories of Poor Households

Type of Household	Fixed capital	Total Capital
Extreme Poor cases	500.0	4500.0
	250.0	1050.0
	200.0	550.0
Average	**316.7**	**2033.3**
Moderate Poor cases	7000.0	7000.0
	3500.0	3500.0
	3800.0	3800.0
	60.0	560.0
	2600.0	2700.0
	300.0	1800.0
	150.0	4150.0
	120.0	920.0
	200.0	3200.0
	2100.0	2300.0
	3500.0	3550.0
Average	**2120.9**	**3043.6**

(Contd.)

(Continued)

Type of Household	Fixed capital	Total Capital
Borderline poor cases	9500.0	9500.0
	13000.0	13000.0
	5300.0	6500.0
	350.0	3150.0
	400.0	10400.0
	300.0	3800.0
	400.0	9400.0
	5000.0	5250.0
Average	**4281.2**	**7618.7**

Source : Case studies, 1996.

A substitute for one's own fixed capital to fund a ME can be provided from supplementary wage earning occupation by a second working member of the family. These earnings can be used to repay a part of the loan and thereby the fixed capital acquired using the borrowed fund can be saved. This linkage demonstrates how a family's human resources may work as a substitute for physical or financial capital.

Another problem for the extreme poor households arises from the fact that most NGOs have a (declared or undeclared) minimum floor for loan sizes. If this amount has to be absorbed as running capital, the ratio of running capital and labour will be such that the marginal product of capital will be very low. Moreover, very poor families may not have adequately secured homestead units where they can shelter what is to them a large amount of money or business capital.

An example will make the problem clear. A peddler who carries head-load of goods and sells them from door to door, cannot carry more than his/her physical capacity. Similarly, a woman engaged in making puffed rice, cannot process more than a certain amount every day. Therefore larger capital cannot be used productively in these activities.

The next question is why cannot hired labour be used to absorb the minimum size of an NGO loan? The first answer to this is that the principle of *self*-employment is envisaged by most NGOs. Even if this is not strictly enforced by most NGOs, the use of hired labour is not practical for most activities pursued by the extreme poor.

Firstly, the marginal product of most non-agricultural activities taken up by the poor is so low that the use of hired labour at the market wage rate will not be profitable. Though there is a prevailing notion that there exists surplus labour in the rural areas of Bangladesh, the nature of the surplus is such that it cannot be hired for non-agricultural employment. Employment in non-agriculture requires a regular year-round involvement, but most wage labourers are employed in crop activities, at least during the peak periods of agriculture. During the periods of peak agricultural activity there is often a scarcity of wage labour. Thus, it is difficult for non-farm enterprises with low marginal productivity to expand further through the use of hired labour.

Another critical constraint in this process is that very poor households may not be enterprising enough to build up a hired labour based enterprise. Nor would the rural wage labourers be enthusiastic about taking up employment with an extreme poor worker starting a new enterprise. After all, the employer needs to give the impression of being economically sound and better off than the wage employees!

As a result of the various constraints described, NGO members usually do not expand in a single activity; they generally prefer to diversify their investment pattern.

The Supply Side Reconsidered

The above discussion puts emphasis on the demand side as a reason behind the restriction of entry into NGO credit by the poorest households. One should, however, reassess whether it is purely due to a lack of demand that the poorest households do not enter as NGO members. One may point to the low profitability of non-farm activities carried out by the poorest households as the reason operating from both the side of supply and demand. The fear that the poorest households may not be able to generate sufficiently high returns from ME investment may discourage the NGOs to provide loans to this group and thus the rate of return becomes a supply side problem.

Even if the demand side plays a crucial role, it may be difficult to remove the constraints which account for low profitability of ME for the poorest households. Still a knowledge of these factors may help the NGOs to adopt proper policies so that NGO credit suits the need of the poorest. Credit NGOs should actively take up the search for means through which credit can reach the poorest and help them to use it profitably so that proper repayment is ensured. A closer look at the operational principles of the NGOs reveal that attention to the following issues are desirable.

Minimum Loan Size

The floor size of loans is often set at around 3000 Taka. Whereas it is true that many poor households complain that the loan sizes are small, there are very poor households which cannot gather sufficient courage to borrow as large an amount as 2000 or 3000 Taka. As has been mentioned, they may not have a sufficiently secured homestead to shelter a cash amount of 3000 Taka or an equivalent investment and they may not possess the human resources for utilizing such a sum. Many of them could never imagine having such a 'large amount' of money in their own hands. For these persons, a loan larger than the capacity to invest will lead to consumption use of the rest of the money, which will then be difficult to repay. Thus all considerations suggest that there should be no minimum floor regarding the size of loan. Borrowers' initial situations should determine this.

Rate of Interest

The effective rate of interest on NGO credit can range from 16 to 50 per cent. A low rate of interest is obviously preferred by the borrowers. It may enable many of the poorest clients to enter the credit market. The problem of reducing the rate of interest is that it cannot be done for only the poorer clients. A uniform reduction of rate of interest will reduce the earnings of the NGO and thus diminish the prospect of their financial sustainability.

For the larger and expanding NGOs, a lowering of the rate of interest may not necessarily reduce their earnings. Interest elasticity of demand for credit may be such that a reduction in the rate of interest will expand the total volume of credit, and raise the number of borrowers per branch. This may enhance the prospect of financial sustainability of the NGO by reducing the cost of the operation and increasing total interest earnings.

Closer Supervision and Monitoring of the Use of Loan

Usually the NGOs engage field level officers to mediate in the process of disbursement of loan and to collect the repayment installments.

They are also expected to help the borrowers in choosing the type of investment. Choice of projects and their actual implementation is usually left to the borrowers. This autonomy may be useful for experienced investors, but for new borrowers from the poorest group, this may lead to a leakage of funds and inefficient use. Closer supervision by NGOs' field level officers may help them to guard against misuse of funds. However, closer supervision may be expensive for the NGOs, because this will require a larger number of staff per borrower. Moreover, many of the experienced borrowers who would borrow large amounts and use a part of the fund for non-investment, would be discouraged from doing so if proper monitoring takes place. This would reduce total disbursement and thus the total interest earnings of an NGO may fall. The way out of this is to adopt different policies for the new, nervous borrowers in contrast to the older, more experienced borrowers. Unlike interest rate differential, this may be easier to implement.

Larger Average Loan Size and Special Loans of Larger Size

This issue is related to the strategy of fixing a minimum loan size. In addition to such a floor, a larger average size of loan may have a positive effect on the prospect of financial sustainability of an NGO. A larger loan size would reduce the cost of operation, since costs per Taka of loan will be lower. Average loan sizes could be increased through various means. For example, the upper loan ceiling may be increased; the usual size of loan granted may be kept closer to ceiling rather than the floor; special schemes granting larger loans to entrepreneurs who expand the scale of activity or who use modern technology could be set up. Such a rise in the size of loans has a bias towards the non-poor. This may not operate directly as an anti-poor bias, but when the total fund is limited, a diversion of large loans to the borderline poor may indirectly imply a bias against the poorest who are thereby deprived.

Even if the larger loans have an indirect anti-poor bias, they are difficult to argue against because of the greater efficiency of the larger loan operation as well as the economies of scale of larger enterprises. To counteract the anti-poverty bias arising from larger loans, the NGOs could make provision for special loans to poor borrowers. The size of these loans would be small, without prejudicing the minimum floor, the repayment period could be longer with some discount on interest rate for early repayment. This might be packaged with adequate supervision to ensure proper utilization.

Concluding Remarks

The previous section suggests a modification to some of the basic features of the lending operations by large NGOs. It is obvious that the NGOs are aware of the desirability of most of these policies. Therefore, one needs to look at the underlying problems of implementing them, in order to seek alternative strategies to pursue these policies.

Most of the policies mentioned above, such as small, flexible loan sizes to suit the poor, longer repayment periods, closer supervision, are likely to raise the cost of credit operation, making it more difficult for the NGOs to reach their goal of financial sustainability. The most important criterion for success of the credit NGOs is their ability to achieve financial sustainability so that they are not eternally dependent on donors for the expansion of their activities. This is desirable of course, but at the same time the NGOs should consider whether the attainment of such sustainability involves too large a cost in terms of the sacrifice of excluding the poorest. The early NGO movement had the objective of poverty alleviation at the forefront and service to the poorest should continue as a criterion of success of the NGOs.

If the aspirations for financial sustainability and the objective of serving the poorest have contradictions, it is likely that the latter will be sacrificed, especially when the donor attitude is to reward the former. In that situation the only alternative that remains for the poorest is the provision of social security. While modifications in credit operations could enable the poorest to join normal economic activities, there can be no justification to push them into a social security programme. If the large and expanding NGOs cannot take up this agenda, the small NGOs may find a meaningful role in this area. The small NGOs often face difficulty in finding clients who are not members of large NGOs. In this situation, if they specifically target the poorest who have been left out by the large NGOs, they could assume a meaningful role. This may be supported by organizations like PKSF through the provision of funds at a lower service charge than the usual. The criterion for success by the small NGOs serving the poorest households should not be financial sustainability, but the extent to which they have reached the poorest and brought them under financial discipline.

Case Studies

Case Study 1: Sharif's Way of Managing Access to Land

This story of a Grameen Bank member shows how a small loan can help a household to come out of poverty and distress and remove a crucial bottleneck in their endeavor to utilize an existing skill.

Sharif is an energetic person in his early thirties, married to Nilufar who is about 22. Sharif worked in their family enterprise of cotton yarn spinning. His father managed the business and he himself along with his three brothers worked in the actual yarn processing. Income from this business was sufficient to maintain the family of five members.

But two years after Sharif's wedding, his father died. Sharif started to manage the business as he was the eldest among the brothers. With his experience and hard work, he felt he could carry on the business successfully. His younger brothers, however, did not take into account the hard realities of life and engaged in extravagant spending without asking Sharif's permission. As a result, the business failed and they were left in debt. Even this downturn in business failed to alert the younger brothers to their responsibilities as partners. At this point, Sharif and Nilufar moved to a separate household and the joint family broke down.

But Sharif did not have any capital to run his business. All his savings and even the proceeds from the sale of a sewing machine owned by Nilufar went to the repayment of debt. They were saved from plunging into further debt when Nilufar became a member of GB and borrowed Taka 3000 to buy a sewing machine. Nilufar's neighbours were members of GB and they provided her with information on this when she was looking for a source of funding for a loan. She bought a sewing machine for Taka 2000 and purchased tools for yarn spinning with the rest of the money. Nilufar earned around Taka 100 per week from sewing, and Sharif started his own business yarn spinning. Unfortunately, due to the small size of capital it was not very profitable. The open space they used for spinning was owned by another person, who charged Taka 100 per month as rent.

Nilufar obtained another loan of Taka 4000 in the next year. They used the money to buy a small piece of homestead land and constructed a small hut which became their home. They chose their homestead land in a location where there was some government owned (*khas*) land nearby, which they could use for their spinning so that they did not need to pay any rent for the land.

At present Sharif brings raw material from traders on credit, processes the yarn and sells it for Taka 10 per kilogram. Their weekly income from this is around Taka 200. The return from this is low because a substantial part of the profit is extracted by the traders. Nilufar plans to take another loan next year so that they can purchase raw materials with that cash and sell the processed yarn directly to the manufacturers. That will more than double their weekly earnings.

The couple have accumulated personal savings of Taka 200 and group-fund saving of Taka 300 with the GB. Such savings are important for self-employed families whose earnings are crucially related to their labour input. It means that they can draw money from their savings in times of emergency.

Case Study 2: Enterprising and Physically Strong Persons are Preferred as NGO Members

Manik Mia and his wife Moslema could not become NGO members. Manik Mia is engaged in petty trading and occasionally in rickshaw plying. He is in his early fifties, his wife in her mid forties; both of them are quite elderly by village standards. That is why he prefers peddling food items and on *haat* (market) days he sits in the market place. Plying a rickshaw requires hard physical labour so he does it only occasionally, when his trading income is barely sufficient for survival. This explains why he cannot borrow from an NGO to buy a rickshaw, as a number of pullers in the locality have done: he would not be able to repay his loan as a part-time rickshaw puller.

Case Study 3: Dependency Burden is a Deterrent to the Proper Use of Loan

Successful NGO members are required to make a proper investment of their loan and to make regular repayments. If the repayment is ensured, it does not matter whether or not it is invested. However, those who do not use their loans for investment are likely to run into difficulties. This happened to Rajab Ali who diverted his loan to consumption. This was done in the face of poverty and the need to feed a large number of mouths. With small children, his wife could not engage in any income generating activity. Moreover, with money in the house (obtained as loans from an NGO), they could not bear to see their children hungry. As the money did not bring in any return, repayments suffered and they were not given any further loans. A landless household with one working member and three dependants is usually among the poorest households, with little scope to improve their situation with NGO loans.

Case Study 4: Success of Female Entrepreneurs

Parbati Rani with her two sisters and mother have to eke out a living for themselves and for her disabled father. She has no skills or education. However she can make puffed rice, which most rural women can do. In addition to this skill, she has had the courage to invest NGO funds in this food processing activity. This generates a sufficient return to repay the loan, maintain the family and reinvest. She has borrowed five times from an NGO. Her business capital is around Taka 4000. She has purchased cows, but she was still unable to get out of poverty because of large health expenditure on her father. However, her condition is assessed as "moderate poor" and the *muri* making business has saved them from going hungry.

An exemplary case of female entrepreneurship is Joytoon who is a widow. Her daughter went to live in her husband's village and her son and daughter-in-law preferred to form a separate household. Joytoon had no property except the house where she lived. She could have taken a job as a maid servant, but instead she took up the business of retailing *sarees* and other clothing. First she started the business with charity from a Union Parishad member. Then she borrowed from an NGO. Now she has a successful business which brings sufficient income to lead a respectable life without depending on the son and daughter-in-law who did not want to support her.

Chapter 14

BREAKING OUT OF THE GHETTO: EMPLOYMENT GENERATION AND CREDIT FOR THE POOR

Geoffrey D Wood

Introduction

Nested within the debates about credit for the poor in Bangladesh is a relatively unexplored issue which relates strongly to purposes and scale of credit, viz. the multiplier effects of credit, especially in relation to employment generation. This is particularly important if there is confirmed evidence that many of the poorest borrowers are dropping out of credit programmes due to the non-viability of their micro-business ventures and consequent non-repayment of loans. Most of the credit models among the NGOs as well as concerned academics seem to assume family based entrepreneurialism as the panacea for 'business-oriented' development, which coincides with modes of provision, repayment schemes, risk perceptions by lenders, assumptions about capacity of borrowers, assumptions about 'what is appropriate' for borrowers, and so on.

Central to these assumptions is that borrowers can all be entrepreneurs, and that this is the route to poverty alleviation and removal. Little consideration is given to employment generation, which seems to be confined beyond the normal production relations in agriculture to either the private sector or public sector works. This represents an over-narrow conception of credit potential when lending to the poor, and also confines credit to non-structural outcomes in terms of more fundamental transformation in the political economy.

There is much emphasis upon group lending, but much less on group or collective productive activity. There is also much emphasis upon the group approach for repayment and credit policing (joint liability etc.). However we need to look more creatively at different models of credit use, involving on-lending, transfers, pooling, collective entrepreneurialism, share-holding which could lead to larger scale, structurally significant lending through which the extreme poor can graduate to less vulnerable positions.

Is it possible to transform the poor of Bangladesh into a nation of small business persons? We accept that a high proportion of educated, literate classes with property and savings work in large organisations and companies as employees. We do not expect or require them to construct livelihoods via their own entrepreneurial initiative. Why, then, do we expect so much of all the poor who are targeted for credit? It seems important to have a more disaggregated understanding of the poor, and to recognise the diversity of their capacities, social position, family circumstances and livelihood options. They are not an homogenous, undifferentiated mass, to be offered an undifferentiated, single, universal package of financial services.

With this obvious premise established, there is a subtle argument to pursue, albeit cautiously. There appears to be a strong case for differentiating the financial product to a far greater extent than at present. This entails moving away from simple versions of the micro-finance formula, which are currently dominating the discourse about credit and the poor. In particular, the range of products should extend from large-scale forms of investment with substantial multiplier effects within local, poverty economies to non-dedicated overdrafts, which enable individual families to manage liquidity and consumption under conditions of high seasonal variation in their income flows. Furthermore, there should be complete transparency about this continuum of options. The latter 'overdraft' option is, in effect, widespread in Bangladesh though not formally acknowledged in the windows through which such credit is accessed. Loans are 'legitimated' by reference to a production plan, but actually this may constitute merely a device for supporting consumption. The Grameen Bank might represent a partial exception to this generalisation, since although it might favour the productive use of credit, it nevertheless retains the notion that 'people know best how to use credit'. However, that proposition is also undermined by the repayment formula, which is insufficiently variegated to address the diversity of liquidity needs.

The argument has to be handled cautiously, because by proposing that the poor be differentiated more widely for financial services there is of course the danger of introducing divisions between the poor, thereby undermining the advantages of solidarity among them. Implied in this division is that not all the poor are equal; that structurally significant credit is only appropriate for some, and 'ghetto-ed' credit for others. The question must eventually then be posed whether the potentially negative features of this differentiation proposition can only be overcome by accompanying social mobilisation? However, there are prior questions to resolve.

Equity and Equality

In what ways are the poor not equal? There is much discussion about reaching the 'hard-core' poor, thus implying an acceptance of the differentiation principle for analytic and targeting purposes. The basis of this differentiation can be conceptualised in terms of resource profiles, underpinning livelihoods. People manage their livelihoods on several 'resource' dimensions: material, human (as in capital), social (as in inclusion or exclusion), cultural (conforming to dominant values, beliefs and valuations in terms of status and honour), political (as in membership of political groupings giving access to other key resources and opportunities), and participation in various forms of co-operative arrangements (e.g. within common property regimes). (See Annex 1 for a more detailed exposition.) Each person or household therefore has a 'resources profile' which reflects virtuous circles or declining spirals in their fortunes. People are distinguished from each other through their distinctive profiles, and in this sense they are unequal. For example, rich people have strong connections to gain access to educational opportunities, good jobs and prestige locations for housing: they are 'rich in people'. Clearly many others have the obverse of these characteristics, and are 'poor in people'. But there are other examples of the profile: under-capitalised; inappropriate skills for the labour market; minority religion; being a women in a strongly patriarchal society; and so on. Embedded within these sets of 'resources' are the socially constructed and culturally conditioned attitudes towards risk-taking as well as knowledge of innovative opportunities in the market, which might lead to at least temporary higher marginal rates of return before others enter and compete in the sector, thereby curtailing monopoly and reducing marginal rates of return on investment. Likewise, within the poor, as defined by poverty lines and other indices of 'quality of life', there is variation in the propensity to take risks as well as knowledge about the extent of risk. On this 'risk' dimension, there is also, therefore, inequality among the poor, perhaps between the entrepreneurial and non-entrepreneurial poor (not necessarily fully overlapping with other classifications such as the declining, coping and improving poor, but close enough).

While programmes of credit for the poor should not set out to exacerbate such inequality, they should avoid a lowest common denominator approach of only designing and targeting credit on those with the weakest profiles: i.e. the extreme poor, since this may not be the best way to reach and assist this category of the poor. The differentiated ability of the poor to absorb different packages of financial services should be recognised. It should also be recognised that the more entrepreneurial poor may provide employment opportunities for the less entrepreneurial. In this way, one might pursue equity objectives without presumptions or objectives of equality by favouring Pareto optimum outcomes in which the risk taking of some provides employment opportunities for others.

Who are the 'Entrepreneurial Poor'?

The concept of the entrepreneurial poor has become pervasive in the micro-finance discourse (albeit, by stealth), but it should be deconstructed. By understanding the 'profile' framework described above, the concept of entrepreneurial poor should be seen more as a socially constructed phenomenon rather than an attitudinal one. In other words, there is a distinction between the 'social' capacity to absorb credit in profitable ways and the 'psychological' ability to do so. Capacity is determined strongly by the structural circumstances applying to the particular actor or household, whereas ability is a more endogenous variable, reflecting personality and skills. Of course the two interact, but the entrepreneurial poor (other expressions are: the 'vulnerable non-poor'; or the 'upper capability' poor) are more likely to be those with more virtuous profiles in which personal psychology is a minor explanant. Once this is accepted, then it follows that the small business approach is not universally appropriate to the target group. Indeed, one might precisely argue that it is most inappropriate for the hard core poor, with the weakest, mutually undermining profiles. In short the prevailing strategy is the least appropriate for hard core poverty alleviation, let alone removal.

Yet, current fashions in development, in this neo-liberal era, are emphasising micro-enterprise as the panacea for poverty alleviation, supported by credit and training. It is reminiscent of earlier rural development populism which focused policy on the small farm sector as an alternative to larger scale farming in which capital and labour would interact through different sets of actors. The problem with emphasising micro-enterprise for strategic support by development agencies to the exclusion of other income earning alternatives (let alone other dimensions of the resources profile) is that it assumes a widespread entrepreneurial propensity among a population whose past survival has been rooted in risk aversion. It assumes a knowledge of markets, and presumes a buoyancy of demand and favourable terms of trade in these micro-enterprise markets. Both assumptions are highly questionable: overriding socialisation, culture and 'peasant rationality' on the one hand; and dismissing the conditions of relative isolation, remoteness and weakness of exit and voice options on the other. Price takers are assumed to be price makers. It is therefore not only necessary to consider where employment options also currently exist, or potentially exist as an object of policy, but also whether there is a linkage between employment and poverty-focused credit interventions.

What are the counter-arguments to such an embracing proposition? First, not all entrepreneurialism is the same. There is a significant difference between: low and high risk activity; familiar and unfamiliar activity; petty trading with low capital requirements and capital demanding productive investment; short and long term turnover. Secondly, it is possible to overcome the profile weakness of the hard core poor by social development interventions such as: group formation to overcome social exclusion; conscientisation; and skill training. The first objection has little teeth since it accepts, in effect, the proposition about differentiated capacities, and then segments economic activity into the dual opposites, but then confines or ghettoises the poor to the former set. (The concept of ghetto credit will be dealt with below.) The second objection is more powerful, and clearly constitutes the basis of critiquing 'credit alone' strategies by insisting that only social development (mobilisation) can create and sustain the conditions (i.e. substitute for weak profiles) under which the poor can deploy credit to longer term structural significance. However the second objection still relies solely upon the small business model, either with individuals supported by a larger collectivity or through collective economic activity itself (i.e. whole group projects), which are more rare (but see a description of such a programme in Wood and Palmer-Jones 1991).

Both these objections, with their corresponding programme content (individualised ghetto credit; or socially supported, ghetto credit; and structurally significant credit) rely upon a degree of social mobilisation in order to reduce the transactions and information costs of institutional lending to the small business objective. But the latter pursues social mobilisation, beyond the motivation of merely reducing lending costs, in order to ensure the social and micro-political conditions under which the long term returns to credit can be secured. In other words, the concept of joint liability extends beyond the institutional objectives of loan repayment to capacity maintenance in the market.

The Problem of 'Ghetto Credit'

What is the 'small business objective'? It seems to constitute economic activity, primarily for family members (or group members in the case of collective projects), for example: petty trading, livestock rearing, and food processing. There are many familiar lists of such activities supported by NGOs and various government and donor supported programmes. They all suffer from two weaknesses: they add actors to already overcrowded trading and petty production markets, thus reducing the returns for all; and they do not generate employment outside the immediate family receiving credit (albeit via groups and peer management). In addition, many of these activities are low turnover, menial-labour intensive and self-exploitative in terms of return on labour input measured by time and calories. Some of these activities are additionally gender and age discriminating, with the more arduous duties undertaken by females and the young (though rickshaw pulling is a notable exception, due to the energy requirements and cultural segmentation of such work—female rickshaw pullers are never seen in Bangladesh). What is the value of such credit to the poor beyond offering short term liquidity and low, even negative value on the marginal productivity of underemployed family labour? It would seem that 'ghetto credit' aptly describes this category of income-generation. This is not to deny its short term 'relief and safety net' contribution towards poverty alleviation, but it also reminds us of its non-structural status in terms of long term poverty removal in a highly unequal political economy. Perhaps its 'employment-generation' equivalent is the cheated labour on public sector, rural works schemes.

Obviously this contrast between 'ghetto' and structurally significant credit has been posed in a deliberately and polemically sharp way. Real life is more shaded than this. There are, of course, many examples of 'graduation' in the sense of an upwardly mobile trajectory, with the extreme poor (often women) working long hours, sacrificing immediate consumption at very low levels of consumption in order to save and reinvest. The case of Saleha (in Hedrick-Wong, Kramsjo and Sabri 1996) is a good illustration of this graduation process, albeit with a sting in the tale (!). But for every one who 'graduates', how many remain on a low level plateau in persistent conditions of vulnerability to minor shocks and disruptions, entrapped in the local, hostile political economy?

Linking Credit and Employment

The preceding discussion therefore makes three principal criticisms of the current micro-credit for the poor discourse: contemporary micro-credit is more attached to poverty alleviation objectives rather than removal ones (with *de facto* relief and safety net functions rather than structural change outcomes); it entails an undifferentiated conception of the target group; and it assumes widespread entrepreneurial capability among the poor attached to expectations of 'graduation'. It is as if there is a mono-culture of micro credit. These criticisms lead to further ones about: modes of credit provision; over-limited conceptions of group liability; insensitive repayment periods; restricted scale; and universalised models of use with limited multiplier effects on the local poverty economy. Many of these issues are discussed elsewhere in this book, but they underpin a concern about the linkage between credit and employment.

The two intervention sectors of income-generation supported by credit and employment generation through public works have operated independently of each other. There has been little consideration of the attachment of employment generation objectives to poverty-focused credit interventions, beyond addressing the under-employment of family members. And here it is important to observe that such notions of under-employment within the family refer much more to the inadequacy of wages or profit margins than to effort or hours. Credit may be a much more **inefficient** way than wage struggles of addressing this problem. Both strategies are trying to assist poor people's entry into markets and to strengthen their capability once within them. However, it does seem pertinent to reflect upon the potential link between credit for the more strongly resource-profiled poor and employment generation for the extreme, weakly resource-profiled, poor; i.e. the possibility of Pareto optimum relations between the entrepreneurial poor and the extreme poor. In this initial stage of the argument, this is less an debate about the 'graduation' of the poor (see Sharif 1996), though there are models of 'graduation' via wage employment schemes rather than micro-credit (see e.g. Sirohy 1996 on India) which are relevant. (These are essentially wage guarantee schemes altering the opportunity cost calculations of other family members to enable investment in both skill acquisition and equipment). However, it is possible to attach 'graduation' ambitions to a differentiated micro-credit process via other institutional arrangements such as share-holding and pooling; see below.

Basically, for the moment, the key question is: can the support for market entry via credit for some actors (i.e. the entrepreneurial poor) provide the immediate conditions for entry into labour markets by others (the extreme poor) at fair and sustainable rates of remuneration? Part of the answer to this question lies within poverty-focused credit and employment policy, and part of the answer in other determinants of economic performance and opportunities within the country and globally.

Significance of Macro Economic Performance and Policy

Considering the latter determinants briefly, the overall growth and dynamism of the economy is clearly crucial to any non-subsidised expansion of employment opportunities, but the same also applies to small business activity supposedly stimulated by micro credit. However the risks attached to larger scale, employment creating investment are higher in the sense that failure entails bankruptcy rather than a staged retreat into self-exploitation; entrapment to others; and consumption sacrifice to meet repayments. For credit financing institutions, the issue is not which outcome is worse for the poor, but which one carries the greater lending risk. A dynamic, opportunity expanding economy reduces (not obviates) the lenders' need for collateral by reducing the risks of failure in a buoyant market. Clearly some of these expanding opportunities are urban, and connected to the performance of global markets and maintenance of comparative advantage (as in the garments industry) which can act as a depressor on local wages. But it is also worth remembering that some globally linked sectors can operate under flexible specialisation conditions, thus reducing the costs of entry and thereby offering opportunities for the entrepreneurial poor (perhaps in collective units) to invest.

But other sets of opportunities would reflect expansion of demand in the domestic market, again urban but of course also rural. Widespread processes of rurbanisation have been stimulated by public sector investment (often donor funded) in roads, embankments, *thana* centres and feeder roads, and of course in the expansion and intensification of agriculture via minor irrigation, especially. The rise of market and *thana* towns offers many entrepreneurial opportunities in production and services activity, not just trading. Micro-credit at the 'ghetto' end of the continuum might capture some of the lower turnover, trading sector opportunities but is unlikely to capture for the poor the more value-added options. This is where the link between structurally significant credit and employment creation can be established.

These 'conditioning' factors certainly contain a clear message for advocacy within the society. While much attention gets focused upon the institutional options for the provision of financial services to the poor, there can be little doubt that the overall performance of the macro economy is the key to any micro success. Thus an element in the linkage between credit and employment creation is lobbying for the appropriate national level policies, conducive to securing the best multipliers on micro forms of investment. This certainly means encouragement for, *inter alia*: flexible specialisation; the wider use of innovative forms of collateral and actuarial assessment of business projects; further de-regulation of bureaucratic controls which have only served to rig the market in favour of those with privileged access and a capacity to incur high up-front transactions costs; improved transportation and communications; tax holidays for entrepreneurial poor ventures (NB leakage problems); removal of subsidised credit rationing; relaxation of collateral requirements to allow third parties to act as guarantors (e.g. banks calling on NGO borrower receivables), accompanied by a tougher response to defaulters with fixed assets collateral; and of course wider access of social lenders to capital markets. Once again, this part of the argument reveals to us the paucity of 'credit alone' strategies, in this case in the absence of appropriate macro-economic advocacy. This is why agencies involved in credit provision for the poor have to be multi-purpose in their activity and objectives in order to secure the full value of the principal aim.

Institutional Options Linking Credit and Employment

Turning to the remaining parts of the answer concerning the role of lending to the entrepreneurial poor as a way of triggering other poor people's entry into labour markets, institutional options for receiving agents may be extended beyond present formulas as a way of increasing both the scale of borrowing and corresponding operational activity leading to employment creation.

One of the central issues is internal group dynamics, as well as the social dynamics between federated organisations (i.e. People's Organisations). All peer monitoring and joint liability is based upon high cohesiveness and shared objectives within the group. But a further assumption is made about the homogeneity of group members in terms of absorptive capacity, needs and psychological orientation with respect to risk. Where criticisms are raised, the usual question posed is whether all group members gain equal access to institutional loan entitlements? However, this model of shared entitlements, or shared on-lending by groups to individuals, can be relaxed in various ways. Intra-group transfers of credit entitlements tend to be resisted by external providers as challenging equity (and the same observation might be applied in the future to inter-group transfers within a People's Organisation [PO] structure). Recently observing Village Organisations within AKRSP in North Pakistan, there is a much wider spread process of intra-group transfer, which is not frowned upon. It is recognised that some group members are more entrepreneurial than others. Non-entrepreneurial individuals thus receive their loan entitlements through the Village Organisation Credit Programme (VOCP), but then decide that other members of the VO could obtain a higher return on capital than themselves—so they on-lend to a fellow group member. In this way, an 'entrepreneurial' member can collect up the loan entitlements of his colleagues (offering them an additional marginal rate of interest ['mark-up']) and make a larger scale investment. The question for Bangladesh is how far these practices exist informally outside the formal knowledge of credit providers.

However, such a process may be unnecessarily complicated. These arrangements have arisen as an 'indigenous' response to the small scale level of credit on offer to the widest possible number in the programme. This occurs because every VO member has an entitlement, as a ratio of personal savings deposited, and they are in effect only borrowing against their own savings as collateral. However, witnessing and tolerating this process as an acknowledgement of entrepreneurial diversity within the VO, AKRSP opened the Medium Term Credit Programme (MTCP) in which an individual VO member can propose an individual, larger-scale, longer gestation project for approval by other VO members for presentation to AKRSP. Such proposals request borrowing beyond the proposers' own level of savings, though rarely beyond the collective savings of all VO members. Thus other VO members stand as collateral for the proposal, and gain a mark up or commission from the proposer for putting their capital at risk. With this principal established, there are many variations of practice. VO members have to calculate the opportunity cost of forgoing borrowing on their own savings while it stands as collateral for a colleague (since the same savings cannot stand as collateral twice in the same time frame). Thus sub-VO combinations may enter such arrangements, where individuals decide that they have no medium term use for such capital (especially if they are employed elsewhere). But also embedded within these arrangements is the prospect of employment of 'dormant' or 'credit-resting' VO members by the 'credit-active' medium term, VO borrower.

Such arrangements, if institutionalised between members over time, start to resemble different forms of shareholding or (if employed within such projects) company share schemes. With such principles and practices occurring between members at the primary group level, it is then possible to scale-up by considering inter-group transfer and pooling arrangements. Again to refer to the AKRSP case, it is clear that there are 'networks of the entrepreneurial poor' operating collectively between groups, pooling their VOCP loan entitlements. In that context, the groups involved may be valley-based clusters, bound together by extended kinship ties, thus offering a traditional institutional vehicle for compliance monitoring. But similar sets of ties link group members in Bangladesh in their immediate locality. So far, these entrepreneurial networks are mainly using loan capital for larger scale trading ventures (especially gaining the advantages of bulk purchasing under problematic conditions of transportation) either in food and other wage goods, or to support other programmes like livestock vaccination, or to manage the export of major cash crops (e.g. apricot). But there is demand and planning for collective, inter-group production and processing projects as well as establishing bus and haulage firms, and agricultural contracting services using tractors and other equipment. Much of this is clearly transferable to Bangladesh. Indeed, there have long been plans in Proshika for establishing sub-regional mechanics' workshop facilities and services, as the logical, federated stage arising from the landless irrigation initiative. All these activities are employment creating for persons other than the primary borrowers initiating the ventures; and the security of these ventures is enhanced by employing fellow, mobilised members who may or may not have a financial stake in the success of the operation, but the certainly have a livelihoods stake as well as a privileged negotiating position with their colleague employers, since both parties are situated in a multi-periodic game rather than a simple, contractual labour market.

Such arrangements also proximate a further practice which makes a virtue out of the heterogeneity and diversity of absorption and other capacities among group members. Different group members or even constituent PO groups can have a diversity of contributions to make to larger-scale enterprises as a reflection of the differentiation described earlier. Some offer business acumen and risk taking propensity, others offer savings for collateral, others offer labour (perhaps in addition), and some may even offer land where small plots remain with the family, or homesteads are passed down through inheritance. Between groups within a PO area, topography and ecology may be a substantial determinant of diversity in group activity, experience, knowledge, profitable activities and so on (e.g. a local level contrast between a single and multiple cropped area), thus prompting the need to integrate groups into the PO via diversity and interdependence, rather than similarity and co-reliance. Thus at both primary group and at other levels of social organisation, a pooling of contributions is likely, which certainly links one element (credit) to another element (employment) in larger scale enterprises. This process also links the observations about differentiation to objectives of 'graduation', in which even the extreme poor can break out of the ghetto through pooling and shareholding arrangements with the entrepreneurial poor.

Annex 1

RESOURCE PROFILES

Each family has a portfolio of resources through which it seeks to arrange survival, enhancement and security of income flows and other services, thus achieving a virtuous circle of mutually reinforcing resources. Vulnerability may therefore be defined as a breakdown of this portfolio, triggering descent into a vicious spiral, in which weakness spreads across the different resources. Such 'resources' crucially include relationships and status rather than just 'things'. People can be poor in assets and human capital, but they can also be poor in the networks which offer access to opportunities, preferential prices, subsidised credit, and so on: i.e. 'people can be poor in people'. This portfolio consists of the following dimensions (which may be enriched by further, locally specific description).

Material: (Assets and Income Flows) The illustrative list offered below also tries to indicate where the main gender responsibilities lie for income flows, as a way of directing the gender content of interventions to support this 'resource'. It is also useful to make a distinction between major and minor markets by gender in order to identify a gender pattern to production and business opportunities.

- agriculture/horticulture   M/F
- livestock   F/M
- nurseries (fruit)   F/M
- plantations (fuel, timber)   M/F
- road construction   M/F
- house construction   M
- business (trading, shops)   M/F
- handicrafts   F/M
- services involving machinery   M
- haulage   M

Human: education, vocational skills, health, low infant mortality, positive gender relations, confidence and emotional security, indigenous capacities, adaptability in problem solving, aspirations, imagination, risk taking, education of parents.

Social: agnatic/affinal ties; inclusion in mosque decisions; clientelism; favour /patronage of stronger, richer families; connections to political, bureaucratic and religious actors; connection to NGO; urban links (education and employment); good marriages; quality of gender relations; involvement in associations and clubs; position in networks.

Cultural: religion, gender (female vulnerability in the context of household and societal patriarchy), customary family status, participation in festivals, inclusion in ritual observance, sports/teams.

Common Property Regime: access/participation in: irrigation, drinking water, rights to barren land, *khas* ponds, electricity, plantations. Inclusion in relevant collective property rights and the rules of their management.

Annex 2

ENTREPRENEURIALISM AND EMPLOYMENT

The following two matrices set out a procedure for identifying the room for manoeuvre in generating income flows between entrepreneurial, private sector employment and public sector employment by material resource/opportunity sector; and by institutional options for intervention and management (with necessarily a higher tautological element). The third matrix shows the actual and potential interaction between resource/opportunity sector and appropriate institutional options. Values can be placed in the 'cells' of the matrix to indicate the intensity of intersection, perhaps on a scale of 1-5 (with 5 the highest). Different participants to a strategic planning process may differ in judgements, and through defence and debate, knowledge of appropriate options is developed. No attempt is made to offer judgements in this paper. Of course, participants may also play around with the categories along the rows and columns to reflect more detail and sensitivity to local context. Thus the term 'community' requires unpacking to precise context. The category 'state' should be broken down into appropriate line departments and special programmes.

Matrix 1: By Resource/Opportunity Sector

Entrepreneurial Poor	Private Sector Employment	Public Sector Employment
Agriculture		
Horticulture		
Forestry		
Livestock		
Water/Irrigation		
Marketing		
Construction		
Processing		
Haulage		

Matrix 2: By Institutional Triangle

	Entrepeneurial Poor	Private Sector Employment	Public Sector Employment
State			
Market			
Community			

Matrix 3: By Resource/Institution

	State	Market	Community
Agriculture			
Horticulture			
Forestry			
Livestock			
Water/Irrigation			
Marketing			
Construction			
Processing			
Haulage			

PART V

MICRO-CREDIT: A RESTRICTED APPROACH TO FINANCIAL SERVICES

Chapter 15

SAVINGS:
FLEXIBLE FINANCIAL SERVICES FOR THE POOR
(AND NOT JUST FOR THE IMPLEMENTING ORGANIZATION)

Graham Wright, Mosharrof Hossain and Stuart Rutherford

"The task Professor Yunus set himself was to show a sceptical world that a formally constituted organization can lend to the poor and expect to get repaid, with interest, and on time. The highly disciplined system he devised therefore paid more attention to guaranteeing the recovery of loans and cutting down on delivery costs than to the banking needs and preferences of his customers. His outstanding success in showing that a very basic loan service can be delivered to the poor has allowed the current generation of critics and practitioners to move on to a search for ways to increase the range and quality of banking services for the poor..."

— Stuart Rutherford, ASA The Biography of an NGO

Abstract

The paper reviews some of the experiences of providing flexible financial services to the poor both in Bangladesh and abroad. In Bangladesh, the large MFIs are using compulsory savings as a source of capital for loans, but there is increasing pressure from the members' for access to their savings. This apparent dilemma may not be real, since experience suggests that open access and other flexible savings facilities may well increase the net savings deposited.

BURO, Tangail has implemented a programme that, in contrast to the large MFIs in Bangladesh, emphasizes savings instead of credit and provides its members with open access to their savings. The study carried out by the authors tried to determine and describe what contribution these savings facilities have made to providing important and valued financial services to the members, and to capitalizing the organization's activities.

The authors conclude that the poor want and need flexible savings facilities. Government authorities and the large MFIs should be less concerned with reducing the rate of interest charged on loans, and more with providing secure, open access saving facilities to the poor. BURO, Tangail's programme suggests that voluntary open access savings *can* raise funds not dissimilar to those levied through the mainstream MFIs' compulsory savings schemes. More research is needed to see if indeed such schemes can attract *even more* savings deposits than the compulsory counterparts.

The paper concludes that this could be the beginning of a new era in Bangladesh when the large MFIs provide a wider range of financial services to a broader spectrum of people and thus improve the indigenous capitalization of their systems. And, noting the risks when less well established MFIs begin savings mobilization, the paper makes a plea for the development of depositor protection schemes.

"The mobilization of voluntary savings is critical for a financial institution that plans to move from an operationally sustainable micro-credit programme to a fully self-sufficient financial intermediary. Since there are not enough government and donor funds to finance global demand for institutional micro-credit, the development of such sustainable organizations is necessary if this demand is to be met. An even more important reason for undertaking the institutional mobilization of voluntary savings is the vast, unmet demand for institutional savings services at the local levels of developing countries worldwide."

— Marguerite Robinson at Microfinance Network,
Cavite, Philippines, November 1995

Introduction

Bangladeshi Micro-Finance Institutions (MFIs) have achieved a reputation worldwide for their success in delivering credit facilities to the rural poor. MFIs are estimated to be reaching 25% of poor households in Bangladesh, offering them credit facilities that in addition to small weekly loan repayments, require a variety of compulsory savings contributions. Nearly 6 million Bangladeshi clients are demonstrating, that "the poor are bankable".

This paper suggests that the MFIs in Bangladesh have not yet fully understood just how "bankable" the poor really are. While they have slowly expanded the range of credit facilities on offer, the major government and non government MFIs have yet to offer high quality savings facilities for the poor. The Bangladeshi large MFIs may be missing an opportunity to provide savings services that the poor want and need while simultaneously finding a valuable source of capital for their credit operations.

The paper has two main sections. The first section examines the experience of savings mobilization abroad and particularly in Bangladesh, and discusses the lessons to be learned. The second section analyses BURO, Tangail's experience with offering open-access saving facilities, and the members' reactions to them, and discusses the challenges and opportunities facing the MFIs in Bangladesh as they start to offer more flexible savings services to their members. The paper concludes by offering some conclusions for policy-makers and programme-managers.

Savings—Abroad

Recently savings have risen to the top of the international development community's agenda. The Consultative Group to Assist the Poorest (CGAP) in its Note 2 of October 1995 stressed "Possibly the greatest challenge in micro-enterprise finance is to expand the provision of savings services to the poor." This is driven by the fact that in the words of Marguerite Robinson, "There is substantial evidence from many parts of the world that: (1) institutional savings services that provide the saver with security, convenience, liquidity and returns, represent a crucial financial service for lower-income clients; and (2) if priced correctly, savings instruments can contribute to institutional self-sufficiency and to wide market coverage.[1]"

Bank Rakyat Indonesia (BRI), with which Marguerite Robinson has worked for over a decade, has mobilized over $ 2.5 billion in voluntary savings through 14.5 million savings accounts and provides

---

[1] Marguerite S. **Robinson**, *Introducing Savings Mobilization in Microfinance Programs: When and How?*, Microfinance Network, Cavite, Philippines, and HIID, USA, November 1995.

services to 30% of Indonesia's households. Contrary to common beliefs BRI *does* attract the very poor: a 1993 BRI study found that over 30% of BRI members had a monthly income of $78 (Tk. 3,120) or less. At any time, for every borrower, there are five savers, and their savings provide the capital for all of BRI's loans. To achieve this extraordinarily level of market penetration and capitalization BRI offers three savings facilities: 1. the liquid which permits unlimited number of withdrawals, 2. the semi-liquid with a restricted number of withdrawals per month, and 3. the fixed deposit. As with traditional banking, these facilities offer clients the opportunity to balance liquidity and returns.

Similar savings facilities have been offered by Kenyan Rural Enterprise Programme, and by BancoSol in Bolivia, with remarkable success.

In the SAARC region, in addition to endless ingenious, indigenous user-owned schemes, the SANASA Thrift and Credit Cooperative Societies in Sri Lanka offer both open access accounts and higher interest term deposits. These have proved successful not only at providing a service that is much in demand and mobilizing savings from which SANASA can make loans, but also in meeting the needs and optimizing the participation of the very poor. Richard Montgomery[2] argues that SANASA's flexible repayment terms, open access savings facilities and instant consumption loans provide vitally important coping mechanisms for the poor in times of stress. Indeed it is this flexibility of financial facilities offered by SANASA that allows it to attract the very poorest as clients.

Savings—At Home

Background and a Glimpse of the Issues

Only recently did the large MFIs in Bangladesh start to use members' savings as a source of financing. Previously savings were simply deposited in formal bank accounts in preference to being lent back to the members.

However the last five years have seen a marked shift. Not only are the larger more established MFIs on-lending their members'

---

[2] Richard **Montgomery**, *Disciplining or protecting the poor? Avoiding the social costs of peer pressure in solidarity group micro-credit schemes*, Working Paper for the conference on Finance Against Poverty, Reading University, March 1995.

savings, but also increasing numbers of smaller NGOs are using members' savings as a key source of capital. Indeed there is evidence that the NGOs mushrooming using this method to start operations are posing a rapidly growing threat to members' savings. Many small NGOs have found that they are unable to meet members' expectation and demand for loans from the small scale savings mobilized. Confidence declines, and unless external capital is found, savings dry up, and the repayments begin to falter...

Despite (or perhaps because of) this experience, there appears to be an increasingly clear consensus that savings are an extremely important issue and a much needed financial service for the poor in Bangladesh. Grameen, BRAC and ASA members have been increasingly vocal about their dissatisfaction with the denial of access to their savings, and many mature members are leaving the organizations in order to realize their (often substantial) compulsory savings. For example in 1995 Khan and Chowdhury noted that nearly 57% of membership discontinuation in BRAC's programme is attributed to the lack of access to group savings during emergencies[3].

The importance the poor attach to savings is also demonstrated by the many (and often costly) ways they find to save. These include investing in assets that can be sold in emergency (for example corrugated iron sheets or livestock), lending between family and friends, or even by taking a loan from an MFI which (Stuart Rutherford argues) are often "advances against savings[4]".

And lest we forget, as far back as 1988, the work of Clarence Maloney and A.B. Sharfuddin Ahmed found that the poor with only Tk. 1,500-1,999 income per month saved 14% of their monthly income, and even those with less than Tk. 500 monthly income saved 2%[5]. Although Hossain Zillur Rahman's article *Times of Hope, Times of Despair* (based on data from Bangladesh Institute of Development Studies' on-going Rural Poverty Trends study) in The Daily Star of July 20, 1996 put "average private investment rate in rural Bangladesh" at 6.7%.

---

[3] KA **Khan** and AMR Chowdhury *Why Do VO Members Drop Out?*, BRAC, Dhaka, 1995

[4] Stuart **Rutherford**, *The Savings of the Poor: Improving Financial Services in Bangladesh*, Binimoy, Dhaka, 1995.

[5] Clarence **Maloney** and A B Sharfuddin Ahmed, Rural Savings and Credit in Bangladesh, University Press, Dhaka, 1988.

LOANS AS ADVANCES AGAINST SAVINGS
(AN IMPORTANT DIGRESSION)

> When Rubhana bought the calf with her first loan, she knew it would be a struggle. Not only would she have to find the Tk. 70 for the weekly repayments, but she would have to buy food for the calf so that it grew and fattened quickly. With a little care with the meagre household budget, and selling the eggs from their few chickens, she felt that she could manage.
>
> And Feroza was confident that, if by the grace of God her husband was well enough to cycle the rickshaw all through the year, she could pay off the loan she had used to buy jewelry for her daughter's wedding, and a few sheets of corrugated iron to replace the leaking thatch on their home. (Of course, she had told the NGO's field worker that she was using the loan for "rickshaw business" to keep him happy.)

Rubhana and Feroza share one thing in common with millions of other MFI members throughout Bangladesh; they are making their weekly loan repayments not from income arising from the loan, but from the normal family household income. This pattern is extremely common not least of all because of the typical MFI repayment schedules. These, largely for reasons of discipline and accounting, require 52 weekly installments (no grace period), and necessitate that the loans be invested in enterprises that generate a rapid rate of return in excess of 100% if repayments are to be made from the enterprise's income.

Thus savings from household income are usually the primary source of the money used to make loan repayments. Thus loans can, and indeed should, be seen as "advances against savings".

Perhaps it is small wonder that organizations offering flexible savings facilities, such as the Credit Unions north of Tongi and SSS in Tangail (see page 18 below), have mobilized remarkable savings (Tk. 350 million ($ 8.75 million) and Tk. 14.3 million ($ 0.4 million) respectively). And it is noteworthy that some of the strongest demand for these savings services come from the poorer clientele. Finally, user-owned savings schemes such as "lottery funds" or ROSCAs are proliferating in Bangladesh suggesting that the informal sector is once again ahead of the semi-formal sector of the MFIs. (See Stuart Rutherford's *Informal Financial Services in Dhaka's Slums in this volume*).

While MFI experts were discussing the poor's ability to save, the poor were sorting out how and where they could save most effectively.

However, to date the large MFIs in Bangladesh have not provided saver-friendly facilities and their massive savings balances have been mobilized by compulsory taxes and levies. An examination of the research and policies of the larger MFIs in Bangladesh reveals issues that may, in part, explain their continued preoccupation with credit-based financial services.

A Detailed Example with Thanks to BRAC

BRAC had a policy that required members to deposit weekly minimum compulsory savings of Tk. 2 (raised to Tk. 5 in 1994), plus 4% of all loans disbursed, into the Group Trust Fund savings account. Thereafter, limited withdrawals were possible (25% after 5 years, 50% after 10 years and 100% after 20 years or on resigning membership) subject to a maximum of 3-5% of total savings balance for each Area Office. This meant in practice that members probably had to leave the organization in order to realize their savings.

In 1993, BRAC introduced a savings experiment in ten branches which were instructed to introduce open access or "current" accounts for the members. However in an attempt to "see different modes of operation and their efficiency[6]" no fixed guidelines were issued from head office as to how the new scheme should be implemented in the field. A variety of schemes were implemented in a variety of branches (aged between 2 and 8 years), and the study concluded that "the average own savings per head was Tk. 17.7 a month compared with Tk. 13.8 in the (matched) control branches." However, as a result of the high level of withdrawals, the study concluded that "in the first year of operation, savings mobilization was not enough to cover the operating expenses[7]".

But these conclusions require careful consideration. In the branches where the savings scheme was genuinely voluntary and open (i.e. where the BRAC staff were promoting the scheme and were not dissuading or blocking members from making withdrawals) the financial results look markedly superior. In Pabna branch for

---

[6] **Hassan Zaman**, Z. Chowdhury and N. Chowdhury, *Current Accounts for the rural Poor. A study on BRAC's Pilot Savings Scheme*, BRAC, Dhaka, July 1994.

[7] **H. Zaman** et al. Ibid.

example the monthly end *net* savings rose ten-fold in the year (as compared to less than three-fold in the comparison control branch at Mohonpur). Furthermore, a look at the detailed financial data suggests that a year may have been too short a window of opportunity for the members... they may have been testing the new savings system to see if it was to operate as advertised prior to depositing, and not withdrawing, larger sums.

Irrespective of any of the above, an indisputable finding of the study was that the members greatly valued the scheme. "The VO members' attitude towards BRAC... have significantly improved due to this new facility" and "The villagers on the whole stressed that they would prefer that BRAC maintain an open savings account even if it meant that lower interest was paid on deposits...[8]"

Indeed, the authors of a follow-up study assert, **not** to allow members open access to their savings "runs counter to the fundamental organizational goal of BRAC[9]"—empowerment of the poor.

But BRAC faces the apparent dilemma now confronting all of the larger MFIs in Bangladesh: "However pertinent it may sound to have a differentiated savings scheme with a provision for complete withdrawal access, at present it is not feasible for BRAC to operate in such a way due to the interdependence of savings and credit programmes, and the operations of the Revolving Loan Fund. In BRAC Rural Credit Programme (RCP) members' savings partially serve as an insurance mechanism against loan default. Furthermore, BRAC relies on members' savings for its Revolving Loan Fund.[10]"

Not only have the Group Funds acted as *de facto* loan guarantee reserves (although this is formally covered through Emergency Fund contributions in many MFIs) but also years of enforced group savings have allowed the larger MFIs to develop a huge capital fund for their lending operations. It is feared that allowing open access withdrawal of those savings could result in massive outflows of funds as the members use these large balances—possibly in preference to taking loans.

BRAC's new liberalized policy on savings withdrawals—introduced January 1, 1996—permits the member (who "must have been

---

[8] H Zaman *et al. Ibid.*

[9] **Farah Deeba** and Ishrat Ara, *A Note on Providing Access to Savings of VO members*, BRAC, Dhaka, November 1995.

[10] **Farah Deeba** and Ishrat Ara, *Ibid.*

a Village Organization member for at least 12 months") to withdraw "any amount of her/his savings". However, "a member will be allowed to withdraw only once in a Bangla calendar year"; and "the withdrawal will be treated as a loan without interest, and must be paid back to her/his account with BRAC within a period of 12 months through monthly (or less frequent) installments".

BRAC has opened another (interest free) credit window from which members can meet emergencies and short term needs, subject to the usual rigorous loan repayment conditions. This will maintain the Group Fund balances and thus contribute to meeting BRAC's capital needs while meeting the members' expressed need for long term saving for their security for old age.

But the new scheme is unlikely to attract those who want to save to buy capital items or fund children's education. Nor those who want open access to their savings in the event of emergencies or opportunities without the prospect of having to add another repayment instalment to their weekly household expenditure. And, ironically perhaps, the member will still have to leave the very Village Organization which promoted her development in order to enjoy her savings in her old age.

One is tempted to suggest that this halfway house is hardly the recipe for massive savings mobilization or for meeting the expressed needs of the members.

Similarly, Grameen Bank

Since its inception, the Grameen Bank system has required a 5% group tax plus Tk. 1 (recently raised to Tk. 2) personal savings contribution to the Group Fund. "To ensure that ownership of the Bank remains in the hands of the poor, and to ensure capital for future growth, it is compulsory for each group to buy shares in Grameen Bank. When the savings in a Group Fund have reached Tk. 600, the group concerned is obliged to buy shares in the amount of Tk. 500 (i.e. 5 shares at Tk. 100 each)[11]." Savings today... share capital tomorrow.

Thereafter, interest free loans are made available from the Group Fund at the discretion of the group. Grameen Bank reports list more than 350 uses of these loans including a variety of social and

---

[11] Andreas **Fuglesang** and Dale Chandler, *Participation as Process—Process as Growth— what we can learn from the Grameen Bank*, Grameen Trust, Dhaka, December 1993.

household needs, health and medical expenses, loan repayment, maintenance, repair and addition to capital equipment, raw materials for manufacturing and processing, farming and trading. And with $ 59.2 million Group Fund loans disbursed from Grameen's inception to December 1995, the facilities have been heavily used by the members.

But until 1995 the members were only allowed to withdraw the personal savings component of the Group Fund and then only when they left their group. The substantial group tax (as much as Tk. 3,000 or more for members who have been with the Grameen Bank for ten or so years) was retained in the group. Only since 1995 have the members been entitled to a full refund (including interest) of Group Fund savings when they leave their groups. And better, after ten years with the Grameen Bank, the members who remain are now entitled to transfer all of their portion of the savings in their Group Fund (with interest at 8.5%) into their Individual Savings Deposits (see below).

Other savings schemes under Grameen Bank's system include the compulsory Emergency and Children's Welfare Funds, the voluntary Special Savings (or Centre Fund) and Individual Savings Deposits designed to encourage the members "to build their economic strength by keeping extra income in personal savings accounts". These Individual Savings Deposits earn 8.5% interest, and may be withdrawn on request (irrespective of whether the member has a loan outstanding) from the Branch Office.

Chen noted back in 1992 that the total savings mobilized by Grameen Bank was four times larger than the combined savings of the five major commercial banks in Bangladesh[12]. And by the end of 1995 Grameen Bank members had generated a cumulative Group Fund savings of $105.4 million, or over 70% of the value of all grants received from donor and other institutions, and were meeting a very important part of Grameen Bank's capital requirements.

The Individual Savings Deposits represent the first open access current account savings scheme among the large mainstream MFIs in Bangladesh. And more, it is policy that people who are not Grameen

---

[12] Noted in Shahid **Khandker**, B. Khalily, Z. Khan, "Is the Grameen Bank Sustainable?", Human Resources Development and Operations Policy Working Paper # 23, World Bank ("findings, interpretations and conclusions expressed... are entirely those of the author(s) and should not be attributed in any manner to the World Bank"), February 1994.

Bank members can also save using these accounts. However, interestingly Grameen Bank's Head Office MIS does not track Individual Savings Deposit accounts, at present they are monitored at Zonal level instead, and are probably subsumed under the Group Fund heading in the Annual Accounts. This may be indicative of a system under development, or the perceived unimportance of these Individual Savings Deposits, (after all, "Grameen Bank has never deviated from its central vision of credit for poverty alleviation[13]"). It also suggests that Grameen may have an under-recognized instrument and capital reserve in these Individual Savings Deposits at the branch level.

With the Grameen Bank's new policy, the Individual Savings Deposits assume a tremendous importance... many members who have been with Grameen for more that ten years are transferring their Group Funds to the Individual Savings Deposits, and some non members are beginning to open accounts.

We examined three sixteen year-old branches[14] in Tangail, with an average of 2,499 members. This limited and rapid survey revealed that an average of 1,451 (58%) of members had Individual Savings Deposits, and on average 110 non-member accounts had been opened in each branch.

Suraz branch, with the highest number of non-member accounts (200), had attracted Tk. 1,858,700—an average of Tk. 1,170 per account. However, the 1,424 members and 25 non-members in Atia branch had only deposited Tk. 461,160—an average of Tk. 318 per account. This low average net savings per member seems to be the result of a high level of withdrawals. In June at the Atia branch, the Tk. 685,153 withdrawals exceeded deposits by 4%, and Grameen suffered a substantial net outflow of funds. The situation was marginally better in Elashin branch where the Tk. 149,834 June withdrawals were only 65% of the savings deposited in the month.

These withdrawals appear to be large scale transactions—in Atia the average withdrawal was Tk. 4,030, a figure that may well correspond to the amount transferred in from the group funds. The relatively large number of accounts, together with the high level of turnover suggests than a significant proportion of members are with-

---

[13] Syed M. **Hashemi** and S.R. Schuler, *Sustainable Banking with the Poor: A Case Study of Grameen Bank*, manuscript under preparation for the World Bank.

[14] Elashin, Atia and Suraz as of 9, July 1996

drawing substantial percentage of their Group Funds once they are transferred into the Individual Savings Deposits. This analysis was confirmed by the Grameen Branch Manager in Atia—almost as soon as the Group Funds are transferred into their Individual Savings Deposits (net of any loans outstanding), the members are withdrawing them.

What is not clear is why these withdrawals are being made. Are the members now using these substantial sums to establish income generating activities without the burden of the annualized weekly interest *and* capital repayments? Or are they being used to improve housing, to pay off burdensome loans, or to buy land and other assets? Or is it to test the new system (which has been introduced after years of member pressure for access to their group fund savings), and with time and confidence, Grameen Bank will see a rise in the level of savings per member? This requires research—it has important implications for the policies and programmes not just of Grameen, but also for the other large MFIs in Bangladesh.

On the basis of their experience in Tangail, there is a possibility that Grameen Bank, like BRAC, may conclude that open access facilities will result in a massive outflow of savings and thus precious capital. But we should not lose sight of the build-up to the change in policy that allowed members to withdraw from their substantial Group Fund savings: a right won after years of protest and, in Tangail, a strike during which many groups did not meet. At the end of 1995, in the whole of Dhaka District there were around 18,000 general loanees with repayments more than 25 weeks overdue, but in Tangail there were nearly 60,000, and the cumulative unrepaid amount had climbed to over Tk. 82 million or $ 2 million. It may take a while for the members' confidence to be rebuilt.

And ASA

In October 1995 ASA modified its mandatory savings policy. Previously, ASA required compulsory weekly savings of Tk. 5-10 (the level is decided by the members), allowing withdrawal of savings only upon discontinuation of membership. Emergency Savings Loans (to a maximum of Tk. 1,000 at 10% pa (APR 20%) interest) were available under a few strictly defined criteria.

ASA's new policy has allowed open access to 80% of the balance of a member's mandatory savings as a loan. The member must repay this loan over one year in weekly installments with 5% (APR 10%) interest, credited to her account. The member must repay one loan-from-savings before taking another.

However ASA have not stopped there. With Save the Children USA, ASA has submitted a project proposal to USAID/Washington to "develop new savings products which are financially sustainable, appropriate and replicable... ASA will open twenty units in various areas of Bangladesh". These units "will offer voluntary savings and credit services to 240,000 clients. Additional savings services will be offered to 24,000 non-members/borrowers who are also expected to benefit from the pilot savings project... ASA expects to offer two types of voluntary deposit accounts, liquid and semi-liquid..."[15] If successful, the pilot savings scheme will be introduced throughout the organisation. The proposal notes that "In addition to providing an important service to the poor, the addition of voluntary savings services will support ASA's further expansion and will reduce the need for additional donor funding".

The proposal is to be implemented soon, probably initially in Comilla, and looks to offer a timely opportunity to explore some key issues facing MFIs in Bangladesh.

Interim Discussion

Despite the increasing recognition of the importance of savings for the poor, until recently all the large MFIs in Bangladesh offered extremely limited savings facilities. Savings were compulsory (and often linked directly to loan size), and members had few, if any, opportunities to withdraw them. It is hardly surprising that most members saw such savings simply as part of the price of a loan.

In view of the demand for access to Group Funds, BRAC and ASA have introduced loans-against-savings schemes very similar in nature to that of the Grameen Bank Group Fund. These loans-against-savings schemes have been developed more on the rationale of the MFIs' institutional necessity to secure and safeguard capital funds than on the members' needs or desires.

But the **basis** of the rationale that limited access to the Group Funds will indeed better secure the capitalization of the MFIs in Bangladesh has not been adequately explored. If the experience

---

[15] Draft of the USAID-Implementation Grant Project to be implemented by Save the Children USA and Association for Social Advancement (ASA), April 1996

from abroad has anything to teach us here in Bangladesh, a wide variety of savings facilities including open access and term deposit accounts may well significantly *increase* the level of net savings deposits and thus capital available to the MFIs. Evidence from the Credit Unions, SSS and others also suggest this might be so.

Grameen Bank has activated its Individual Savings Deposits, BRAC examined Current Accounts briefly in 1993, and ASA is to start its own flexible savings facilities pilot scheme soon. The large MFIs attach great importance to the well-being and needs of their members, and to ways of assisting the poorest 15% of the population generally acknowledged to be unwilling or unable to join most MFIs' schemes. They will, no doubt, continue to seek new ways of offering financial services to address these needs... and indeed the increasing competition among them also necessitates this.

We could be at the beginning of an important new era in which the large MFIs in Bangladesh introduce flexible savings facilities.

Marguerite Robinson sounds a note of cautious optimism on adding voluntary savings to a micro-credit program: it will "change the entire program. The institution should be prepared for this, and should not think that it is adding just another product. Sustainable micro-finance programs that offer loans and well-designed savings instruments have many more deposit accounts than loans. ...This pattern occurs primarily because most micro-finance clients want to save all the time, while most want to borrow only some of the time."

And there are problems that must be addressed sooner rather than later. At present under the credit-based systems, the MFI must "select borrowers who are trusted by the lending institution. In savings mobilization however, it is the customers who trust the institution[16]." The unfortunate experiences of Social Economic Development Organization (SEDO) and Bangladesh Unemployed Rehabilitation Organization (BURO) in the late 1980s, and of dozens of small NGOs throughout the country over the last six years, dictate rapid development of schemes to protect savers, and help the poor identify reputable solid institutions[17].

In this context, it is perhaps ironic that we should turn to BURO, Tangail, which rose from the wreckage of BURO, for information on

---

[16] Marguerite S. **Robinson**, *Ibid*.

[17] For a discussion of issues arising from this see Stuart **Rutherford**, *Regulation of savings and credit NGOs in Bangladesh: what is it and why is it needed?*, Binimoy, Dhaka, April 1996.

how more flexible savings facilities meet the needs of the members, and how they might affect the capitalization of an organization. Nonetheless, BURO, Tangail is now operating what is adjudged to be a largely successful[18] system that is almost typical of most the MFIs operating in Bangladesh. It has, however, one important difference that makes it worthy of more careful attention: flexible savings facilities.

Description of BURO, Tangail's System

Background

BURO, Tangail initiated its work in 1989 with the establishment of five "model" branches which were the basis of its project development system. The process of on-going monitoring led to improvements in the services and operation of the organization during the small-scale experimentation phase that lasted until November 1991. However, it was not until 1993 that BURO, Tangail had attracted enough donor capital funding[19] to initiate full-scale activities.

By the end of December 1995, 20 branches were operating. The ten branches established in the experimentation phase were turning average profits of around $ 2,000 a month. The other branches were also turning small-scale profits. All profits were re-invested in the branches' revolving loan funds. As of June 30, 1996 donors had contributed $ 332,067 to capitalize the organization, and the 24,000 members had more than matched this with $ 173,578 from branch profits, and $ 214,508 in their savings accounts.

BURO, Tangail continues to conduct operations research to increase the flexibility and responsiveness of its financial services.

Savings Deposits

BURO, Tangail has a minimum weekly compulsory savings level of Tk. 5, but members are welcome to deposit more. Prior to January 1, 1996, members could deposit up to a maximum of Tk. 50 at the group meetings. The Mid-Term Review team, examining BURO, Tangail's operations in July 1995 noted that this policy was meant to

---

[18] Tahrunnesa **Abdullah**, Stuart Rutherford, and Iftehkar Hossain, *BURO, Tangail—Rural Savings and Credit Program—Mid Term Review: Final Report*, Dhaka, 1995.

[19] AusAID, CIDA, ODA, PACT/PRIP, and SIDA

discourage "non-target groups" from joining and in fear of encountering cash management problems. But it was causing problems for savers, and meant that members could not deposit "lumpy" household income from the sale of assets: livestock, housing materials, or from windfalls.

In response to this observation, BURO, Tangail introduced a policy of offering the facility of unlimited savings deposits effective January 1, 1996. Members are permitted to deposit to a maximum of Tk. 200 at the group meeting, and an unlimited amount at the branch. Operations research is underway to examine a Fixed Term Deposit Scheme—offering higher rates of interest for longer term savings—which was introduced on a pilot basis in March, 1996.

Savings Withdrawals

Since its inception in 1989, BURO, Tangail has operated a policy of offering open-access savings accounts to its members. Members are permitted to withdraw their savings whenever they do not have a loan outstanding; and loan taking is entirely optional, at the member's discretion.

BURO, Tangail has taken the decision to test a completely open savings withdrawal system under which members may withdraw at any time. However, the savings balance for the loan size sanctioned must remain on deposit. This condition is designed to increase the members' "equity" and "self-reliance" over time. The table below outlines the savings balance required for each loan size.

Loan Size (Tk.)	Required Savings Balance (Tk.)
< 5,001	15% of Loan
5,001-10,000	20% of Loan
10,001-15,000	25% of Loan

Nonetheless, neither the system currently in operation (limiting withdrawals to when members have no loans outstanding) nor the proposed system (requiring members to maintain required savings balances to receive loans) satisfy Marguerite Robinson's recommendation to provide completely open access savings facilities unlinked to the members' credit status in order to maximize savings deposits.

The Aims of the Study

The study aims to determine and describe what contribution the BURO, Tangail savings facilities have made to:
   a. providing important and valued financial services to the members, and
   b. capitalizing the organization's activities.

Methods

Quantitative Survey

A questionnaire was developed in discussion with BURO, Tangail management, and field tested by three VDWs who interviewed 15 members. After a discussion group to de-brief the VDWs, the questionnaire was modified on the basis of the field test.

The revised questionnaire was given to a sample of 241 members drawn at random from 120 *kendras* or "centres" under two old branches (Silimpur and Pathorail—100 members) and one new branch (Elanga—41 members). The questionnaire was administered by six/seven VDWs who received half a day of training on the questionnaire and approaches to administering it.

Data was entered in Tangail, and cleaned and analysed in Dhaka. Incomplete or anomalous data was checked to the questionnaire form, and back to the members through re-interview where necessary.

Qualitative Survey

The qualitative survey team carried out 16 interviews of thirty minutes to two hours duration. Some interviewees were selected on the basis of their responses to the quantitative questionnaires and others at random. Most interviews were with individuals or very small groups of women, but some were with men, and two were with bigger groups. Our main aim was to learn what is meant by savings, who does it, how and where, what is its significance in the household economy, its history and growth rates, and how does it relate to NGOs and to their loan programmes.

Limitations

This study was by no means comprehensive, and raises many unanswered questions. The sample size is too small to draw any detailed conclusions from the data, and particularly data disaggregated over length of membership, or by income band. Nonetheless, broad themes and issues were clearly and easily identifiable from the data gathered and interviews held.

Results and Discussion—Quantitative Survey (Summary Analysis attached)

Demographic Details

The typical member in the sample is 33 years old (although the ages range from 19 to 68), she is married and wife of the head of the household (86%), and has been a member of BURO, Tangail between 1 and 7 years (average 3.3). 45% of the members have grade 1-5 education, while another 44% have no education at all. On average 5 family members are eating from the same rice bowl (range 2-12 family members), and typically two of the family members are earning income. This is a fairly representative socio-demographic profile of Bangladeshi NGO MFI customers.

Income

The Tk. 3,677 average monthly income per family is higher than the median reflecting some high earning families (around 15 over 10,000 per month). The median monthly incomes are around Tk. 3,000 in Pathorail and Silimpur, and Tk 2,000 in Elanga (total range Tk. 1,000-20,000)—though we should bear in mind the "notorious under reporting of income". These incomes are also typical for Bangladeshi MFI customers, suggesting that the open access savings facilities have not, to date, been successful in attracting large numbers of the poorest 15% of the population that the large MFIs are generally acknowledged to be missing[20]. Neither of the surveys were able to explain this omission, although there was some suggestion that the very poor are concerned even by the Tk. 5 per week compulsory savings requirement.

The members have a broad range of income sources. 36% are involved in cow/goat/poultry rearing, another 13% in farming land and another 15% in weaving/handicrafts. Although 20 members (6.6%)

---

[20] The Mid Term Evaluation Team that looked at BURO, Tangail in July 1995 reached similar conclusions.

used their loans for informal money lending, only 6 reported informal money lending as a source of income.

Loans

Ninety-three per cent of the members have taken a loan, and on average they have taken 2.2 loans (range 0-6). The average size of the most recent loan is Tk. 4,019, (and the median about the same). 17% of loans were used for cow/goat/poultry rearing, another 16% for land purchase/mortgage and 15% for weaving/handicraft, and 15% for small business or trade/shop.

It is interesting to note that although only about half of BURO, Tangail's members are taking loans at any one time, nearly all of them have used the credit facilities at one stage or other. Few members have elected (or been able?) to use BURO, Tangail just to establish a current account. If we accept the argument that the very poor want and need to use savings-only facilities, this begs some the same questions as the apparent lack of the very poor among BURO, Tangail's members.

Savings Now

On average, the members surveyed had Tk. 867 in their savings accounts at the time of interview in the first two weeks of June, an increase of 61% on a year before. But this hides differences amongst the branches: Elanga's newer members had an average of Tk. 541, but members in Pathorail had Tk. 1,141 and in Silimpur, members had Tk. 726. Pathorail members have saved 24% more and withdrawn 30% less than the Silimpur members. Median savings balances in all the branches were between 10 and 18% lower than average savings which were dragged up by a few large balances.

Over a comparable period, BRAC would have probably raised a total of Tk. 1,145[21] per member, Grameen would have probably levied a total of Tk. 730[22] per member, and ASA members would have probably saved an average of Tk. 1,238[23]. Thus BURO, Tangail's system has raised 19% more than that of Grameen, but only 76% of that of BRAC, and 70% of that of ASA.

---

[21] 3 loans totalling Tk. 8,000 @ 4% plus Tk. 5 per week for 3.3 years compulsory savings.

[22] 3 loans totalling Tk. 8,000 @ 5% plus Tk. 2 per week for 3.3 years compulsory savings.

[23] Tk. 7.5 per week for 3.3 years compulsory savings

Savings During the Year to 31 May, 1996

On average, the members had each deposited Tk. 455 savings in the year to 31 May, 1996, and 22% of them had made withdrawals—particularly in Pathorail and Silimpur. This resulted in an average net savings per member of Tk. 333. This is 27% up on the average net savings per annum over period of membership (Tk. 262) and suggests a growing confidence in BURO, Tangail and its system (not entirely surprising since the organization was under-capitalized, unable to meet the members' demand for loans until late 1993, and operating in Tangail where almost every villager has heard of an NGO that has set up shop, collected savings, made a few loans and then disappeared). The average withdrawal in the year to 31 May, 1996 was Tk. 507.

The limited data available, suggests that there is no significant increase in the members' annual savings over time, that is to say that the members in the year to 31 May 1996, typically (median) saved around Tk. 350 (net) a year or Tk. 7 a week—whether they have been with BURO, Tangail for one or seven years. The survey data suggests that the number withdrawals may reduce, and the Tk. amount withdrawn increase, with length of membership. The net effect appears to be a declining level of withdrawals over time, a trend confirmed by analysis of BURO, Tangail's consolidated annual accounts 1992-1995 (see below).

The average weekly gross savings figure is raised to Tk. 9 per week by a few "big savers" and particularly active accounts. This average is a very creditable Tk. 4 (80%) top up of the Tk. 5 per week compulsory saving requirement. These voluntary savings are providing the annual capital equivalent of the 5% group tax on a Tk. 4,000 loan under the Grameen system. However, when the 27% of annual savings withdrawn are also factored in, this drops to a Tk. 2 (40%) top up of the compulsory saving requirement. Nonetheless, the net Tk. 333 per member savings per annum can make a good contribution to the capitalization of loans... particularly if only half the savers want to borrow.

In a year, from a member that had joined 3.3 years before, BRAC would probably raise around Tk. 410[24], Grameen would probably levy around Tk. 300[25]. and ASA's members would have probably a

---

[24] Tk. 4,000 loan @ 4% plus Tk. 5 per week.

[25] Tk. 4,000 loan @ 5% plus Tk. 2 per week.

contributed an average of Tk. 375[26]. Thus the survey results suggest BURO, Tangail's system is raising 11% more savings than that of Grameen, but only 81% of that of BRAC, and 89% of ASA. Nonetheless, these figures also indicate that as the members gain confidence, BURO, Tangail's voluntary open access savings accounts are beginning to raise similar amounts of capital as the compulsory savings schemes of the large MFIs.

Savings Over Time

Examination of BURO, Tangail's consolidated annual accounts figures suggests a steady level of around Tk. 300 annual savings per member over the period 1992-95, and the annual withdrawals per member declining from Tk. 205 (72% of savings deposits) in 1992 to Tk. 100 (33% of savings deposits) in 1995. Nonetheless these figures give net savings figures for the year to 31 December 1995 of only Tk. 202 compared with the survey's members who net savings figures of Tk. 333 for the year to 31 May 1996. However, (assuming that there are no problems with the sampling-frame), much of this difference is likely to be attributable to the many new members joining in late 1995 (inflating the denominator and thus reducing the per member figures), and perhaps a little to further improved figures in 1996 when BURO, Tangail introduced its new open deposit system and was one of the few NGOs to maintain services throughout the political unrest.

Comparisons with the large MFIs are also difficult because of the rapid growth in membership amongst these organizations . . . consolidated per member comparisons are distorted by the dynamics of members joining over the years.

Nonetheless trying to make adjustments for this (by taking the average of the prior year membership and the current year membership), the attached comparison shows that BRAC and ASA's figures are similarly reduced over the calculations above—probably as a result of the demography of membership. BRAC's annual net savings per member grew from Tk. 157 in 1993 to Tk. 233 in 1995, and in 1995 the comparable figure for ASA was Tk. 264. Thus in 1995 BURO, Tangail's net savings per member was about 80% of that of BRAC and ASA... confirming the survey's figures.

---

[26] Tk. 7.5 per week.

Perhaps we should look at the proportion of loans outstanding are funded by savings in 1995, BURO, Tangail: 37%, BRAC 39% and ASA 40%. Or in case there are significant idle funds, at the proportion of total capital derived from savings in 1995, BURO, Tangail: 27%, BRAC: 25% and ASA 40%. And just to put it all in perspective, the total savings generated from members as of 31 December, 1995. BURO, Tangail: Tk. 0.6 crore, or $ 0.2 million, (Tk. 443 per member) BRAC: 80 crore or $ 20 million (Tk. 559 per member), and ASA: Tk. 24.5 crore or $ 6.1 million, (Tk. 731 per member).[27]

Society for Social Service (SSS)

As another small scale digression to illustrate the market for savings... the medium size NGO SSS also offers open access savings accounts (the only requirement is that members save a compulsory Tk. 5 per week, and maintain 25% of the balance of any loans taken), and is attracting substantial savings deposits. The Mogra Branch Manager showed accounts that indicated in a good week the members saved an average of Tk. 16 per head, and in a bad (floodtime) week they saved an average of Tk. 8 per head. But SSS's 1995 annual audited accounts show the weekly deposit per member averaged around Tk. 6, and that the weekly withdrawal per member averaged only Tk. 1. Since inception in 1990, with PKSF funds at its disposal, SSS has mobilized net savings of Tk. 1.1 crore or $ 0.3 million from 27,183 members—Tk. 628 per member on the basis of 1994/95 average adjusted membership.

Since joining BURO, Tangail, 48% of the 241 members surveyed had made 186 (range 0-7 per member) withdrawals averaging Tk. 415 (range Tk. 56 to Tk. 2,480), and their most recent withdrawals averaged Tk. 460. The highest withdrawal (Tk. 2,480) was to pay service charges to send a member of the family abroad to work. It also appears that the average withdrawals are getting larger but less frequent over time—perhaps reflecting the completion of trial withdrawals and the members' growing confidence in BURO, Tangail. At its present level of activity, the open access withdrawal facility seems unlikely to be making great demands on the staff's time. But we should note that open (and thus irregular) savings deposits are likely to necessitate more time-consuming accounting systems.

---

[27] Grameen Bank's audited financial statements for 1995 were not available at the time of this draft: 29 July 1996.

The more typical uses of the members' most recent savings withdrawal can be grouped into "quality of life" uses (household consumption, improving housing and health care—38%), investments (acquiring land, expansion of business activity and education—31%) and social (marriage and other functions and acquiring gold/jewelry—16%). Only 7% used the withdrawal to make a loan repayment—a concern cited in the BRAC savings studies. In contrast the members planned to use their savings in a markedly different manner. 50% of the members planned to use savings for social purposes, 30% for investments, and 13% did not know. Evidence of reality shattering dreams?

Before joining BURO, Tangail 68% of members had some cash savings—of these, 45% of members had savings in the household, and another 36% had savings with another NGO. 62% of members could withdraw, and 28% received interest on, their savings at an average rate of 44% (range 0-200%).

At the time of interview, 50% of members had savings in the addition to their savings with BURO, Tangail. Of these 33% had saving in the household, another 34% had savings with another NGO, and another 12% had lent them out with interest. 61% of members could withdraw, and 52% received interest on their savings at an average rate of 40% (range 0-125%).

Either BURO, Tangail's current savings facilities are not meeting all the savings needs of the poor, and/or the members are spreading their risk by saving in several places. It could be both of these, and the other savings outside BURO, Tangail may simply be men's savings. In many cases, some household savings are held by the husband, but in other cases some savings may being used as a source of immediate access cash to overcome BURO, Tangail's stipulation that those member with loans outstanding cannot withdraw. Certainly Stuart Rutherford found many ASA members with around Tk. 900 household savings to call on "in case of need"[28].

Results and Discussion—Qualitative Survey

Perceptions of Savings

"*Loans are for development, savings for future security—especially that of the children*".

---
[28] Stuart **Rutherford**, *ASA—The Autobiography of an NGO*, ASA, Dhaka, 1995.

The qualitative survey showed that for women, saving (making *shonchoi*) is now *very* closely identified with membership of an NGO, to the point of being almost synonymous with it. Other forms of saving such as saving at home in bamboo or clay "banks" or by forming user-owned *samities* are now either discontinued or seen as trivial or second-best—the temptation to spend savings kept at home was repeatedly cited as a problem. Virtually all older women confirmed that before the NGOs came they saved at home (and sometimes in local *samities*). Some younger women said that before the NGOs came they did not think about or understand savings and that it is the NGOs that instructed them.

> Shamima, Rokeya and Anwora all agreed, they had always made savings by putting aside a few taka from the sale of eggs and fruit—after all "a good wife saves". The savings had been useful for school fees, clothes and other small-scale expenses. Before they had kept their savings in clay money banks—shaped like *hilsha* fish—but there was always the temptation not to put money in, or to take some out for little treats for the children or visiting guests. But Rubia and Nurjahan disagreed, they had not saved before, Rubia because she "had no place to save", and Nurjahan, because she "did not think about saving". They all liked BURO, Tangail's savings account system, the ability to withdraw their savings and the discipline of having to save each month. "Now we save more, much more", they chorused.

Virtually every women denied that they or their fellow-members "save only to get loans". Several said they joined BURO, Tangail to save and were slowly persuaded that borrowing would be useful.

Considerable sensitivity to interest rates was expressed by many respondents, and they are aware of the different rates. Several pointed out that NGO rates on savings fall well below what can be obtained in the village from lending out. There is a good case for a MFI experimenting with a relatively high rate on a liquid savings scheme, even if that means maintaining loan rates at a high level. In such a scheme, more members might save and fewer borrow: good for them as it would lower their costs and their stress levels, and raise their self-reliance, good for the MFI because it would earn more from fewer loans.

Gender Issues

Savings was and largely still is seen as a female activity. As such its *source* is associated with traditionally female occupations such as poultry raising or other *bari*-based production, and the *use* of savings is often associated with female-controlled work such as child rearing: the most common use cited for savings was schooling costs. This view is shared by men, women and school children. One group of women laughed outright at the suggestion that men could save out of their earnings. This may reflect the men's partiality to tea and cigarettes, films and cards, or may mean that saving is still seen as a small-scale matter. Before the NGOs came women saved at home in sums up to Tk. 500 or perhaps Tk. 1,000. Now, they do not appear to be unhappy with the fairly small size of their NGO savings balances. However there is some suggestion that this may be changing.

Using savings as more than a store for trivial amounts of "female" money is giving way (in some households) to using savings as part of a household-wide strategy for economic growth. This represents a challenge and an opportunity for often female-focused MFIs to attract potentially larger "male" savings.

> Hashimon is a poor BURO, Tangail member from Sit Kazipur village. She and her husband have agreed on a plan. She saves heavily out of energetically-pursued poultry keeping and, though they have two small sons, she saves from 30 to 50 taka a week regularly and has built up a large balance. She has taken two loans (Tk. 2,500 and 4,000) which he uses for trading in rickshaw parts and he repays these from his normal income from rickshaw riding. The idea is that they will soon stop taking loans and rely instead on savings regularly drawn down on and replenished. She has an unusual record as a saver—once, before the NGOs, she saved as much as Tk. 3,000 over three years at home in a clay bank. Her neighbours in this very poor-looking *bari* see her as something of a role model and are proud of her. They told me she is "doing something new". They do not see her as a miser, she says, "rather they praise me".

The Bigger Picture

It is hard not to feel optimistic when carrying out interviews in villages close to Tangail town. There is a sense that attitudes towards economic growth are changing. Whereas before poor people felt that their economic status was fixed, they now have seen and believe in the possibility of personal economic growth, and are busy pursuing it. We were told that men are less idle and more willing to strive for assets or income, and (over and over again) that women are more aware of the role of their NGO memberships in this.

> Minu smiled confidently and assured us that "everyone" is saving more these days, and that this is "part of the progress and development in the whole area". People are busy buying cows, raising poultry and working hard—almost everyone is in an NGO group.

Whether NGO financial services programmes sparked this off or arrived in its wake is not clear—but they are certainly playing an important role in it.

Concluding Discussion

BRAC's concern that "savings mobilization was not enough to cover the operating expenses[29]" is being borne out by BURO, Tangail's experience to date... the additional costs (such as they are) of operating open access savings accounts have not yet been covered by a massive in-flow of capital for on-lending. But the trends suggest that as loans become available and the members' confidence in BURO, Tangail's systems grows, the net savings deposited are growing steadily.

We must not forget that while BURO, Tangail's system is more flexible than the loans-against-savings systems of the other large MFIs, it is still not the most responsive and friendly of savings facilities. More open access accounts are now being introduced on a trial basis. But the accounts studied were still primarily derived from compulsory savings, "minimum balance for loans" and "no withdrawals by loanees" policies. And no term deposit facilities are offered to meet the poor's longer-term needs.

As we are reminded by Marguerite Robinson, "Voluntary savings contrast sharply with compulsory savings required as a condition for credit; these reflect two different underlying philosophies. The latter assumes that the clients must be taught financial discipline and "the savings habit". The former... assumes that most of the working poor

---
[29] H. **Zaman** et al. Ibid.

already save, and that what is required for effective savings mobilization is for the institution to learn how to provide instruments and services that are appropriate for local demand[30]."

Evidence from home and abroad from the remotest of Bangladeshi villages to Dhaka's slums suggests that the poor want to save, and indeed *are* saving in a wide variety of ways. In contrast to the cities where traditional user-owned savings schemes are burgeoning, in the villages around Tangail MFIs have a near monopoly on female saving. What has been lacking until very recently among most MFIs is the facilities to allow the poor to save in a way that they can meet current needs and opportunities as well as save for the future. The large MFIs have instead concentrated on providing credit facilities at the lowest sustainable interest rates, and on capturing compulsory savings in order to do so.

Given this demand for flexible savings facilities, and the fact it appears that most people (including the very poor) want to save, while not all want to borrow all the time, is it not reasonable to suggest that the large MFIs should look at optimizing the savings facilities they offer, even if this entails marginally increasing the rate of interest charged on the loans? Such a policy could allow the MFIs to offer services to a larger number of the poor, and encourage the participation of the very poor in the MFI's programmes.

In addition, with many MFIs focusing almost exclusively on serving the female side of the savings market, the male sector is, as yet, largely untapped. Attracting voluntary male savings, either directly (possibly through outlets in a *haats*[31]), or indirectly through female MFI members, should be an attractive option for female-focused MFIs.

The latter approach might offer opportunities to strengthen the position of women in the family. At least one study[32] has suggested that women believe that their sitting in the weekly (time-consuming) groups to undertake financial transactions is viewed as valuable and important work, and as a consequence, their husbands treat them with more respect and love. Some commentators would put this down to (*inter alia*) the fact that the women were handling money, a

---

[30] Marguerite S. Robinson, *Ibid*.

[31] For a discussion of this see Stuart **Rutherford**, *Why ASA Should Become A Money Management Centre For The Poor*, A note. Dhaka, 1996.

[32] Graham A. N. **Wright** and Shahnaz Ahmed, *"A Comparative Study of Urban Savings and Credit Non Government Organizations in Bangladesh"*, Dhaka, 1992.

traditionally male activity in Bangladesh. If men begin to save as part of a larger household-wide financial strategy, and do so through the intermediation of women, it would represent an important social and economic shift.

The next few years may well prove that if the large MFIs offer genuinely open access savings and other term deposit savings facilities, both the members, and the implementing organization will benefit. Meeting both female *and* male members' needs for a safe home for their savings could also meet a growing portion of the MFIs' need for capital. On the face of it, it appears to be a donor's dream, and is certainly worth exploring in depth.

Finally, even if we are willing to accept that there is a large scale of demand for open access savings facilities in Bangladesh, are MFIs in Bangladesh *ready* to start mobilizing voluntary savings from the public? In terms of the history and performance of many of the MFIs in Bangladesh, they are unquestionably organizationally capable, if not ready. However there are two other conditions that Marguerite Robinson notes as "dominating the issue of *when* microfinance programs should start mobilizing voluntary savings from the public". The first is an enabling macro-economy, an appropriate legal and regulatory environment and a reasonable level of political stability, and the second is effective supervision of institutions providing micro-finance.

Which returns us to the theme of the need for the MFIs in Bangladesh to be proactive in developing regulatory mechanisms and depositor protection schemes...

Summary Conclusions

1. In Tangail "savings" have become almost synonymous with NGO-MFIs' activities, which have replaced traditional forms of saving. The MFIs have generated large-scale deposits with savings facilities that are generally compulsory and restricted in nature. Despite this, there is clear consensus that **the poor want and need flexible savings facilities** to meet emergencies and protect assets, to acquire assets and develop their businesses, and to meet life cycle/social obligations such as education, marriage ceremonies and funeral expenses.
2. Given the poor's need for flexible savings facilities, and particularly if we accept that in many cases loans are indeed sim-

ply "advances against savings", it is reasonable to suggest that **government authorities and the large MFIs should be less concerned with reducing the rate of interest charged on loans, and more with providing secure, open access saving facilities to the poor**.

3. The programmes of BURO, Tangail suggest that voluntary **open access savings *can* raise funds not dissimilar to those levied through the mainstream MFIs' compulsory savings schemes**. and do so while offering an important service to the poor. Nonetheless, it also appears from BURO, Tangail's work from 1991-96 that:

    a. the poor (quite rightly) take time to develop confidence in an organization and its ability to repay savings—they like to see loans being made available on demand and to test the withdrawal facility to ensure that it operates as advertised. Banking is all about confidence.

    b. the very poor (for whatever reasons) have not become members to date.

    However, BURO, Tangail's programme is not adequately established to see if *even more* savings will be mobilized as the members gain confidence in the system, or as it becomes more fully open access. An important challenge for BURO, Tangail, and others, is to attract savings from men too, and there are signs that this is beginning as men deposit increasing amounts into their wives' accounts. It is also still unclear whether the very poor would join a fully open access savings scheme (with no weekly compulsory saving requirement) if invited and encouraged to do so. These issues have important implications for the organization's interest rate structure, and for its capital financing requirements and should be addressed in the next few years.

4. **More research is needed** on Grameen Bank's Individual Savings Deposits, and those committed to the development and improvement of MFIs in Bangladesh should pay careful attention to the operations research programmes being implemented by ASA and BURO, Tangail.

5. This could be the beginning of a new era in Bangladesh when the large MFIs provide a wider range of financial services to a

broader spectrum of people and thus improve the indigenous capitalization of their systems. However, we note that:

    a. the small NGOs mushrooming to take savings and lend them back to members almost invariably rapidly run into a demand for credit in excess of the deposits they are able to attract. This results in declining confidence in the organization until either it collapses or it finds donor/soft loan support to capitalize its credit operations and regenerate confidence.

    b. given the exposure to this catastrophic loss of confidence, fraud and/or incompetence, only those organizations with a track record of delivering effective credit services should be encouraged to move into large-scale savings mobilization.

6. And in this context, lest we forget the anguish and heartbreak of losing two years' of savings in the roar of a departing bus ... **let us together make a commitment to develop depositor protection schemes!**

Comparative Summary of Financial Services Operations for BURO, Tangail, BRAC and ASA for 1995

	BURO, Tangail	BRAC	ASA
Membership ('000)			
Reported Membership	20.9	1,511	404
Adjusted Membership	14.7	1,435	336
Receipts & Payments (Tk. '000)			
Annual Savings	4,432	370,000	88,679
Annual Withdrawals	1,465	36,000	0
Annual Net Savings	2,967	334,000	88,679
Annual Savings per member	301	258	264
Annual Withdrawals per member	100	25	0
Annual Net Savings per member	202	233	264
Balance Sheet (Tk. '000)			
Loans Outstanding	17,673	2,047,898	609,101
Savings	6,509	802,000	245,631
Emergency and Other Funds	1,177	51,119	35,480
Grant RLF	12,335	2,377,533	355,793
Loan RLF	–	0	88,580
Retained Profits	4,367	9,575	23,766
Savings/Loans Outstanding	37%	39%	40%
Savings/Total Capital Funds	27%	25%	33%
Savings per member	443	559	731
Grant & Loan RLF per member	500	1,607	968
Savings/RLF per member	75%	35%	75%

MEMBERS QUESTIONAIRE

Summary Analysis

No.	Particulars	Elanga (1)	Elanga (2)	Pathorail (1)	Pathorail (2)	Slimipur (1)	Slimipur (2)	Summary (1)	Summary (2)
1.	Branch Name								
2.	Kendra No	29	45	46	120				
3.	Name of Member	41	100	100	241				
4.	Savings Account No.	41	100	100					
5.	Number of Years she has been a member of BURO, Tangail	1.4	3.8	3.5	3.3				
6.	VDW's estimate of Member's age	30	35	32	33				
7.	Matrial Status								
	A. Unmarried	1	2.4%	2	2.0%	1	1.0%	4	1.7%
	B. Married	38	92.7%	94	94.0%	76	76.0%	208	86.3%
	C. Widow	1	2.4%	1	1.0%	18	18.0%	20	8.3%
	D. Divorced	1	2.4%	3	3.0%	2	2.0%	6	2.5%
	E. Deserted/Separated	0	0.0%	0	0.0%	3	3.0%	3	1.2%

(Contd.)

(Continued)

No.	Particulars	Elanga (1)	Elanga (2)	Pathorail (1)	Pathorail (2)	Slimipur (1)	Slimipur (2)	Summary (1)	Summary (2)
8	Education level								
	A. LA	0	0.0%	0	0.0%	0	0.0%	0	0.0%
	B. HSC pass	0	0.0%	1	1.0%	0	0.0%	1	0.4%
	C. SSC pass	1	2.4%	3	3.0%	0	0.0%	4	1.7%
	D. Grades 6 - 10	7	17.1%	6	6.0%	9	9.0%	22	9.1%
	E. Grades 1 - 5	16	39.0%	37	37.0%	55	55.0%	108	44.8%
	F. No education	17	41.5%	53	53.0%	36	36.0%	106	44.0%
9	Relationship of interviewee to head of the household								
	A. Wife	35	85.4%	88	88.0%	86	86.0%	209	86.7%
	B. Daughter	3	7.3%	5	5.0%	6	6.0%	14	5.8%
	C. Daughter - in - law	2	4.9%	2	2.0%	5	5.0%	9	3.7%
	D. Niece	0	0.0%	1	1.0%	0	0.0%	1	0.4%
	E. Aunt	0	0.0%	1	1.0%	0	0.0%	1	0.4%
	F. None - is head of household	1	2.4%	3	3.0%	3	3.0%	7	2.9%
	Economic Indicators								
10	Number of family members (eating from same rice bowl)	188	4.6	515	5.2	533	5.3	1,236	5.1
11	Number of income earning family members	58	1.4	164	1.6	247	2.5	469	1.9

(Contd.)

(Continued)

No.	Particulars	Elanga (1)	Elanga (2)	Pathorail (1)	Pathorail (2)	Slimipur (1)	Slimipur (2)	Summary (1)	Summary (2)
12	Approximate gross Tk. income of family for last month	96,400	2,351	340,300	3,403	449,369	4,494	886,069	3,677
13	What are your sources of income?								
A.	Paddy husking	5	5.1%	1	0.5%	4	1.2%	10	1.6%
B.	Puffed rice	1	1.0%	0	0.0%	1	0.3%	2	0.3%
C.	Cow rearing	5	5.1%	14	7.7%	38	11.6%	57	8.8%
D.	Goat rearing	6	6.1%	7	3.8%	36	10.9%	49	7.2%
E.	Poultry rearing	27	27.3%	22	12.1%	77	23.4%	126	19.4%
F.	Vegetable production/trading	13	13.1%	12	6.6%	32	9.7%	57	9.0%
G.	Fishing/ Fish trading	3	3.0%	4	2.2%	2	0.6%	9	1.7%
H.	Rickshaw/rickshaw van	3	3.0%	11	6.0%	11	3.3%	25	4.4%
I.	Handicraft	4	4.0%	9	4.9%	13	4.0%	26	4.4%
J.	Small business or trade/shop	8	8.1%	23	12.6%	18	5.5%	49	8.9%
K.	Weaving	2	2.0%	34	18.7%	22	6.7%	58	10.9%
L.	Farming land	15	15.2%	18	9.9%	50	15.2%	83	13.0%
M.	Sweet meat production	0	0.0%	1	0.5%	0	0.0%	1	0.2%
N.	Money lending	1	1.0%	2	1.1%	3	0.9%	6	1.0%

(Contd.)

(Continued)

No.	Particulars	Elarga (1)	Elarga (2)	Patrorail (1)	Patrorail (2)	Slimipur (1)	Slimipur (2)	Summary (1)	Summary (2)
O.	Butcher	0	0.0%	4	2.2%	0	0.0%	4	0.9%
P.	Service	6	6.1%	6	3.3%	8	2.4%	20	3.4%
Q.	Other	0	0.0%	14	7.7%	14	4.3%	28	5.0%
		99		182		329		610	

Loan and Business Activities

14	How many loans has the member taken from BURO, Tangail since she joined the organization?	58	1.4	267	2.7	204	2	529	2.2
15	What was the Tk. size of her current/most recent BURO, Tangail loan?	130.500	3,183	459,500	4,595	378.500	3,785	968.500	4,019

16 What was the uses of her current/most recent BURO, Tangail loan?

A.	Paddy husking	9	14.5%	5	3.9%	0	0.0%	14	4.1%
B.	Puffed ride	0	0.0%	0	0.0%	6	5.2%	6	2.2%
C.	Cow rearing	3	4.8%	13	10.2%	0	0.0%	16	5.0%
D.	Goat rearing	2	3.2%	2	1.6%	13	11.3%	17	5.9%
E.	Poultry rearing	11	17.7%	2	1.6%	2	1.7%	15	4.4%
F.	Vegetable production/trading	3	4.8%	3	2.3%	4	3.5%	10	3.2%
G.	Fishing/ Fish trading	3	4.8%	3	2.3%	1	0.9%	7	2.2%

(Contd.)

(Continued)

No.	Particulars	Elanga (1)	Elanga (2)	Pathorail (1)	Pathorail (2)	Slimipur (1)	Slimipur (2)	Summary (1)	Summary (2)
H.	Rickshaw/rickshaw van	1	1.6%	5	3.9%	1	0.9%	7	2.3%
I.	Handicraft	3	4.8%	4	3.1%	6	5.2%	13	4.3%
J.	Small business or trade/shop	7	11.3%	19	14.8%	21	18.3%	47	15.7%
K.	Weaving	0	0.0%	17	13.3%	15	13.0%	32	10.9%
L.	Land purchaes	4	6.5%	11	8.6%	13	11.3%	28	9.4%
M.	Land Mortgage	5	8.1%	6	4.7%	10	8.7%	21	6.9%
N.	Irrigation/tubewells	0	0.0%	2	1.6%	1	0.9%	3	1.0%
O.	Sweet meat production	0	0.0%	0	0.0%	0	0.0%	0	0.0%
P.	Household consumption (Food, Fuel, Clothing etc.)	2	3.2%	4	3.1%	1	0.9%	7	2.2%
Q.	Marriage or other social functions	0	0.0%	2	1.6%	1	0.9%	3	1.0%
R.	Improving housing	4	6.5%	6	4.7%	9	7.8%	19	6.3%
S.	Health care	1	1.6%	1	0.8%	1	0.9%	3	1.0%
T.	Education	0	0.0%	0	0.0%	0	0.0%	0	0.0%
U.	Informal money lending	1	1.6%	15	11.7%	4	3.5%	20	6.6%
V.	Other	3	4.8%	8	6.3%	6	5.2%	17	5.6%
				128		115		243	

Savings Activities
Savings Inside BURO, Tangail

No.	Particulars	Elanga (1)	Elanga (2)	Pathorail (1)	Pathorail (2)	Slimipur (1)	Slimipur (2)	Summary (1)	Summary (2)
17.	What levels of savings accounts are being maintained (Taka)								
	A. Now	22,164	541	114,124	1,141	72,649	726	208,937	867
	B. One year ago	14,980	365	72,137	721	42,550	426	129,667	538
	C. Before member joined BURO, Tangail	5,350	130	53,080	531	15,109	151	73,539	305
18.	During the period June 1, 1995-May 31, 1996								
	A. Total savings deposited (Tk)	17,660	431	51,814	518	40,156	402	109,630	455
	B. Total amount of withdrawls made (Tk)	3,055	764	14,219	508	12,119	466	29,393	507
	C. Number of withdrawls made (#)	4	9.8%	28	27.0%	26	21.0%	58	21.6%
19.	Since the number joined BURO, Tangail								
	A. Total savings deposited (Tk)	25,994	634	141,042	1,410	113,349	1,133	280,385	1,163
	B. Total amount of withdrawls made (Tk)	3,845	549	28,167	397	40,231	373	72,243	413
	C. Number of withdrawls made (#)	7	12.2%	71	53.0%	108	57.0%	186	47.7%
20.	What was the Tk. amount of the most recent savings withdrawls made?	3,245	649	23,742	485	22,895	402	49,882	478

(Contd.)

(Continued)

No.	Particulars	Elanga (1)	Elanga (2)	Pathorail (1)	Pathorail (2)	Slimipur (1)	Slimipur (2)	Summary (1)	Summary (2)
21.	What were the uses of the most recent savings withdrawls made?								
	A. Household consumption (food, fuel, clothing etc)	1	20.0%	6	10.3%	11	19.3%	18	15.7%
	B. Marriage or other social functions	0	0.0%	6	10.3%	1	1.8%	7	5.0%
	C. Improving housing	0	0.0%	2	3.4%	8	14.0%	10	7.3%
	D. Health	1	20.0%	8	13.8%	8	14.0%	17	14.9%
	E. Education	0	0.0%	6	10.3%	4	7.0%	10	7.2%
	F. Acquiring land	2	40.0%	3	5.2%	0	0.0%	5	9.0%
	G. Acquiring other tangible assets (specify)	0	0.0%	0	0.0%	0	0.0%	0	0.0%
	H. Expansion business activity	0	0.0%	15	25.9%	6	10.5%	21	15.1%
	I. Acquiring gold/Jewelry	1	20.0%	3	5.2%	7	12.3%	11	10.6%
	J. Other (specify)	0	0.0%	9	15.5%	12	21.1%	21	15.2%
		5		58		57		120	
22.	What does the member plan to do with her current savings?								
	A. Household consumption (food, fuel, clothing etc)	0	0.0%	2	1.7%	3	2.9%	5	1.9%
	B. Marriage or other social functions	5	8.9%	30	25.9%	20	19.2%	55	20.2%
	C. Improving housing	6	10.7%	10	8.6%	2	1.9%	18	6.2%
	D. Health care	0	0.0%	0	0.0%	1	1.0%	1	0.4%
	E. Education	4	7.1%	6	5.2%	6	5.8%	16	5.8%

(Contd.)

(Continued)

No.	Particulars	Elanga (1)	Elanga (2)	Pathorail (1)	Pathorail (2)	Slimipur (1)	Slimipur (2)	Summary (1)	Summary (2)
F.	Acquiring land	17	30.4%	12	10.3%	16	15.4%	45	15.8%
G.	Acquiring other tangible assets (specify)	0	0.0%	0	0.0%	1	1.0%	1	0.4%
H.	Expansion business activity	1	1.8%	8	6.9%	12	11.5%	21	8.0%
I.	Acquiring gold/Jewelry	14	25.0%	32	27.6%	30	28.8%	76	27.7%
J.	Other (specify)	1	1.8%	0	0.0%	0	0.0%	1	0.3%
K.	Does not know	8	14.3%	16	13.8%	13	12.5%	37	13.3%
		56		116		104		276	

Savings Outside BURO, Tangail

23. If member cash savings before she joining BURO, Tangail where were they held

A.	In the household	9	31.0%	38	45.2%	35	50.7%	82	45.1%
B.	With relatives	2	6.9%	2	2.4%	0	0.0%	4	2.2%
C.	With Friends	0	0.0%	1	1.2%	0	0.0%	1	0.5%
D.	With employer	0	0.0%	1	1.2%	1	1.4%	2	1.1%
E.	With another NGO	13	44.8%	27	32.1%	25	36.2%	65	36.0%
F.	With a formal sector bank/post office	2	6.9%	8	9.5%	2	2.9%	12	6.3%
G.	As informal "rin" loans	1	3.4%	5	6.0%	5	7.2%	11	6.1%
H.	As informal "howlat" loans	2	6.9%	2	2.4%	1	1.4%	5	2.8%
I.	Other (specify)	0	0.0%	0	0.0%	0	0.0%	0	0.0%
		29		84		69		182	

(Contd.)

(Continued)

No.	Particulars	Elanga (1)	Elanga (2)	Pathorail (1)	Pathorail (2)	Slimipur (1)	Slimipur (2)	Summary (1)	Summary (2)
24.	If member had cash savings before she joined BURO, Tangail was the member able to withdraw the savings?								
	A. Yes	17	73.9%	39	50.6%	45	69.2%	101	62.3%
	B. No	6	26.1%	38	49.4%	20	30.8%	64	37.7%
	Total Members able to withdraw their savings	23	56.1%	77	77.0%	65	65.0%	165	68.5%
25.	If member had cash savings before she joined BURO, Tangail, did she receive interest on the savings?								
	A. Yes	5	21.7%	29	37.7%	14	21.5%	48	28.3%
	B. No	18	78.3%	48	62.3%	51	78.5%	117	71.7%
26.	If yes what was the interest rate?	36.0%		40.6%		49.9%		43.7%	
27.	Does the member still have cash savings in addition to her BURO, Tangail account?								
	A. Yes	21	51.2%	64	64.0%	36	36.0%	121	50.2%
	B. No	20	48.8%	36	36.0%	64	64.0%	120	49.8%
28.	If she does, where are they held?								
	A. In the household	11	47.8%	20	26.0%	12	33.3%	43	32.7%
	B. With relatives	2	8.7%	5	6.5%	2	5.6%	9	6.5%
	C. With friends	0	0.0%	0	0.0%	1	2.8%	1	1.2%
	D. With employer	0	0.0%	1	1.3%	0	0.0%	1	0.5%
	E. With another NGO	6	26.1%	34	44.2%	10	27.8%	50	34.3%

(Contd.)

(Continued)

No.	Particulars	Elanga (1)	Elanga (2)	Pathorail (1)	Pathorail (2)	Slimipur (1)	Slimipur (2)	Summary (1)	Summary (2)
	F. With a formal sector bank/post office	1	4.3%	7	9.1%	2	5.6%	10	6.8%
	G. As informal sector "tin" loans	2	8.7%	8	10.4%	5	13.9%	15	11.6%
	H. As informal "howlat" loans	1	4.3%	2	2.6%	1	2.8%	4	3.0%
	I. Other (specify)	0	0.0%	0	0.0%	3	8.3%	3	3.5%
		23		77		36		136	
29.	If member still has cash savings in addition to her BURO-Tangail account does she receive interest on the savings?								
	A. Yes	5	23.8%	43	67.2%	17	47.2%	65	51.5%
	B. No	16	76.2%	21	32.8%	19	52.8%	56	48.5%
30.	If yes what is the interest rate?	54.8%		28.5%		45.1%		39.8%	
31.	Can the member withdraw her other cash savings?								
	A. Yes	18	85.7%	20	31.3%	29	80.6%	67	61.0%
	B. No	3	14.3%	44	68.8%	7	19.4%	54	39.0%

Chapter 16

INFORMAL FINANCIAL SERVICES IN DHAKA'S SLUMS

Stuart Rutherford

Abstract

The spotlight of public attention has shone so fiercely on Bangladesh's credit-giving NGOs that user-owned and other informal devices have received little attention. This paper describes some of the wealth, variety, complexity, strengths and weaknesses of informal finance '*samities*' in some Dhaka slums. Then it examines what these schemes tell us about the demand for financial services among the urban poor, and it goes on to speculate about what NGOs might learn from all this. The conclusion reached is that the urban poor feel a pressing need for financial services that help them manage their cash resources—above all their capacity to save—on a day-to-day basis. This suggests that NGOs would do well to reconsider whether they have been right to emphasise productive loans so strongly in their programmes.

The Schemes

When Clarence Maloney and Sharfuddin Ahmed wrote their handbook *Rural Savings and Credit in Bangladesh* in the mid-1980s they found only two rotating savings and credit associations (ROSCAs), and concluded that '...in Bangladesh the concept (of ROSCAs)... is scarcely known, and there is no understanding of the usefulness of this method of spontaneous group financing' (Maloney and Sharfuddin: 105).

I arrived in Bangladesh at about that time. I had worked with ROSCAs in other countries and I knew from the literature that they are common in the sub-continent. But I too made a disappointingly fruitless search for ROSCAs in Bangladesh.

How different is the case today! The ROSCA, locally known as the *loteri samity*, *khela samity*[1] or *serial*, can be found in cities, towns, country bazaars and even in villages all over Bangladesh. In some areas—such as the slums of Dhaka—their occurrence is extremely dense, with many poor and middle-class households involved in two or three at a time.

And yet when I talk to people involved in NGO-sponsored savings and credit schemes about *loteri samities* I still get blank looks from some[2]. So I had better start with a short illustration.

Mrs Khadeza's Loteri *Samity*

Mrs Khadeza and thirteen of her neighbours in the Aziz Mahalla area of Mohammadpur have formed a fourteen-women loteri samity. They meet weekly when each contributes a hundred taka to their samity. At each meeting one of the fourteen, chosen by lottery, takes the 'prize'— the total contributed that week, in this case 1,400 taka. Once a member has received her prize her name cannot be entered in the lottery again, though she goes on depositing her hundred taka each week. After fourteen weeks every member has had her prize. The loteri samity ends. Mrs Khadeza and her fellow members usually then start another one, perhaps with an extra member or two, or with a bigger contribution and consequently a bigger prize. Meanwhile, Mrs Khadeza is also a member of two other loteri samities, both of them run by local shopkeepers.

Four themes that are important to my argument have already emerged from what I have written so far. The first is that devices like *loteri samities* are essentially ways of *intermediating household*

---

[1] Literally, 'lottery club' or 'game club'.

[2] It seems they are easy to overlook. A recently published book about life in Dhaka's slums was sub-titled *'Dynamism in the Life of Agargaon Squatters'*. Despite the fact that *loteri samities* are one of the most prevalent examples of dynamism in Agargaon they get not a single mention in the book's 104 pages.

savings into useful lumps of capital. The second is the way in which *loteri samities* are *time-bound*: they end after a known period during which all the members have both contributed savings and benefited from the prize. Thirdly we note that in a *loteri samity* the savings deposited are standardised and regular—this *disciplined* form of saving is an essential ingredient of most successful user-owned financial services schemes. Lastly, note that Mrs Khadeza is a member of not one but of three *loteri samities*: *multiple access* to user-owned schemes is common, especially in the urban areas.

In this paper I shall use the term ROSCA rather than local names, because not all ROSCAs use the lottery to decide who gets the prize. Some decide it by consensus and others—though this is still rare in Bangladesh—by auction.

ROSCAs are one of two broad categories of user-owned savings and credit schemes. In the other category the fund built up out of member savings is not rotated among the members as soon as it is created, but accumulates and is there for members (and sometimes non-members) to draw on by taking loans, or not, as they wish. Such schemes are therefore sometimes called ASCAs (*accumulating savings and credit associations* [Bouman]) though I am going to refer to them as 'fund *samities*' in this paper. Let us take a simple example.

Halima Begum's 'Fund'-Type *Samity*

Halima Begum lives in Mirpur One and is a member of the fifteen-person 'fund' *samity* run by her neighbour Mrs Nurjahan. Halima's husband is a daily-paid casual labourer but the couple is ambitious and desperately wants to get him a regular job—a *chakri*. They think this will cost around 20,000 taka. So Halima saves 100 taka a month in the *samity*, which started in December 1995. It will run for exactly one year. Members are allowed to take loans and Halima has taken a loan of 1,000 taka, used to pay back a previous more expensive loan from another source. On her *samity* loan she pays five taka per hundred per month interest, and she will be obliged to return the principle before the *samity* ends in December 1996. In December she expects to get her savings back plus a dividend of about 150 taka—this being her share of the interest earned on loans. This expectation is based on the fact that she earned a similar dividend in a similar *samity* run by Mrs Nurjahan in 1995. Halima also belongs to a local ROSCA.

Note that this 'fund' *samity* is, like a ROSCA, time-bound. It will last just one year. This annual closing of the accounts will serve as a practical audit, and will give a clear view of whether the 'fund' has been properly run or not. Mrs Nurjahan is running her sixth successive annual 'fund' *samity*, and all have ended successfully so far. In this way she has built up a group of loyal members like Halima. Again like the ROSCA, the 'fund' *samity* intermediates savings into loans and finally into lump sums for the savers after one year, and is disciplined, relying on set deposits at equal and known intervals. And again, multiple membership is common: Halima belongs to a ROSCA as well, and Mrs Nurjahan, who runs the 'fund' *samity*, has run many in the past and is running three at present. Mrs Nurjahan is also a member of a Proshika *samity*, though she had been running her own *samities* long before she had heard of Proshika.

The Study Sample

Since January 1996 I have been studying such *samities* in some of Dhaka's slums—mainly in the Mohammadpur, Agargaon and Mirpur 1 areas. Altogether my four assistants[3] and I have made enquiries about ninety-five *samities*[4], using a simple question sheet during interviews with *samity* members met by chance during door-to-door walk-abouts.

When I began I intended to collect careful figures on the frequency of *samities* by type of *samity*. I had assumed—wrongly—that I wouldn't find many *samities*, and I wanted data on which types were becoming established. I quickly found that there were more *samities* than I could count in almost every neighbourhood I explored. Because many people belong to more than one *samity* at once, and because many *samities* are short-lived by design, the task of working out exactly how many *samities* there are in a neighbourhood at any one time proved beyond me. Among the ninety-five *samities* I came across in the three areas there were slightly more 'funds' than ROSCAs (53% were 'funds', 44% were

---

[3] Three women and one man.

[4] We talked to ninety-five respondents, and from each respondent took down the details of *one samity* she or he was then a member of. Since many respondents are or have in the past been members of several *samities* we got, in addition to 95 sets of data, incidental information about dozens of other *samities*.

ROSCAs, and 3% were other types). But because the average life of a ROSCA is shorter than that of a 'fund' it may be that more people experience a ROSCA in any one year than experience a 'fund'.

Among all types, mixed-sex *samities* are the most common: 54% have both men and women. Of the single-sex *samities* in our sample male ones are five times commoner than female ones. Very often *samities* include children or teenagers.

Almost 70% of *samities* that I saw were described to me by their members as being made up of people living in the same neighbourhood. Of the rest, all but one[5] were based on the workplace. But then I and my assistants did our research by walking around the *bustees* during day-time, and the number of *samities* based in garment factories and other work-places is probably under-represented in our small sample.

We restricted our research to neighbourhoods we would describe as 'lower-middle class or poor', and then asked our assistants to assess the economic status of the members they talked to. They described two-thirds of *samities* as being made up mainly of 'poor' people (54%), or of 'very poor' (10%). The other third were described either as 'lower middle' (28%) or 'middle' (8%). To put this subjective assessment into perspective, the NGO-formed group members that we also came across fell into these categories in roughly similar proportions.

The oldest 'fund *samity*' that we found had started in July 1976 and there were eighteen that had been formed before the end of 1994. The oldest current ROSCA started in October 1994, but there were several examples of ROSCAs that are the successors of a string of ROSCAs run by the same leader and going back many years.

The smallest ROSCA we found was a five-person one (among youngsters in a garment factory) and the smallest 'fund' a nine-person all-women one. The biggest ROSCA in our sample has 122 members[6], but there were eighteen 'funds' with a membership of more than 100. Of these, the biggest two had around 500 each: mostly, these big 'funds' are based in garments factories.

By far the most popular way to save in these schemes is daily—two thirds of all *samities* of both types require their members to save

---

[5] That last one was described as based on family relationships.

[6] We had a report of one with 150 members but the respondent wasn't very reliable.

every day. A little over a quarter save monthly—reflecting the presence of people with monthly wage packets, as in many of the garment factory *samities*. Less than 3% use the weekly saving frequency so common in NGO schemes. Virtually every *samity*, of whatever type, requires a fixed[7] savings amount, though some allow members to choose among several contribution 'bands' at the outset of the scheme. Of all savings regimes the single most common is ten taka a day—found among 'funds' and ROSCAs alike. But the range is large: in 'funds' in garments factories we found members saving 100, 150 or 200 taka monthly, whereas in the smaller neighbourhood 'funds' 1, 2, 5 or 10 taka a day is the rule.

Among ROSCAs, the 'prize' (the size of which is a function of savings size, savings frequency, draw frequency and member numbers) ranges from 1,000 to 12,000 taka and our samples were fairly evenly distributed along that range. In the 'funds', the single most common interest rate on loans to members is 10% a month—more than half of our sample have chosen that rate. The lowest reported rate was 3% p.m. and the highest 20%. Most *samities* let their members decide what to do with the loans they take, but some rickshaw 'funds' and several rickshaw ROSCAs insist that members use their loans or prizes to buy rickshaws. Almost half of the 'funds' allow non-members to borrow, but they often impose a requirement for physical collateral on such borrowers, and/or charge them a higher interest rate. About half of all 'funds' say they will distribute the profits made on loans as dividends to members at the end of one year: others intend to run for two or three or even five years before closing the books.

Finally, 14% of all respondents said that there are or have been problems of some sort in their *samity*, usually difficulties in depositing savings or loan repayment instalments[8], or dissatisfaction with the way the *samity* was being run. The others either said there were no problems or that they were not aware of any. Many said they trusted the *samity* leadership because it had run successful *samities* previously, and this was true in some cases where respondents had said there were problems. A general comment was that problems are becoming less frequent as time goes by.

---

[7] That is, each and every member is required to save the same set amount each time a savings is made.

[8] Our research period included the anarchic politics of February-March 1996.

Twelve Representative Schemes

Out of the ninety-five *samities* we found, I selected twelve representative schemes which I looked at in more detail, by taking interviews from leaders and from several randomly-picked members and (in most cases) repeating them over time, to get an idea of how well the *samity* is running. Of the twelve, five are ROSCAs, and five are 'fund' *samities*, and one is a combined ROSCA-plus-fund type. Between them they exemplify a wide range of schemes, since some are male, some female and some mixed, some are running well and some badly, some are based in the neighbourhood and others in the work-place (including garments factories and rickshaw garages), some are new and some are long-established, and the number of members ranges from a handful up to several hundred.

The twelfth scheme is neither a ROSCA nor a 'fund' type, but a co-operatively-owned business based on an initial subscription from over a hundred *samity* members. I include it to remind us that there are other savings-based informal *samity* types in Dhaka: I could as well have described an insurance scheme that I found in West Agargaon, for example, or a 'building society' in which members save only to buy land and build homes. One of the twelve is now a 'fund' *samity* in name only: it has been captured by *mastans* and has become a front for a *chanda* collecting racket[9].

Because of space restrictions I have not described all twelve of these *samities* in this paper.

The ROSCAs

We begin our discussion of the detailed investigations with the ROSCAs. ROSCAs are known throughout the world, and their virtues have been pretty well documented by now [Bouman, Ardener et al]. They are perhaps the most *efficient* form of financial intermediation known to mankind, since they turn small savings into loans instantaneously without any paperwork or storage costs. They are flexible, and can be arranged to suit almost any group—from nineteenth century subsistence farmers in Vietnam who drew prizes in a rice-based ROSCA once per agricultural season, to modern Japanese businessmen keen to get capital without the bureaucracy of the formal banking system. The life of the ROSCA, the deposit amount,

---

[9] Illegal collection of 'subscriptions'; protection racket.

the period of the draw, and the number of members, are all variables that can be chosen to suit circumstances and then changed at the end of the cycle if those circumstances change[10]. The lottery system used by most Dhaka ROSCAs is seen as fair, since every member is treated alike. Where some members are more interested in saving but others are keen to get a lump sum of capital quickly, the order in which the 'prize' is taken can be decided by consensus, or (even better) by auction, a device which rewards savers for holding back and taking the prize late in the cycle, turning the ROSCA into a lucrative investment opportunity[11]. ROSCAs are transparent, suiting illiterate members who can see each meeting who is contributing and who is receiving. Finally, ROSCAs can be fun.

The ROSCA, it seems, came late to Bangladesh. What have Bangladeshis made of it? Judging by what I have seen, they are in the process of making a huge success of it, and along the way they are adding several innovations, tailor-made to local conditions.

We have already described Mrs Khadeza's *loteri samity*. I chose it because it exemplifies the classic basic ROSCA, found worldwide. But it isn't typical of Dhaka ROSCAs. The typical Dhaka ROSCA is more likely to be a rickshaw ROSCA like those run by Nurul Haque, or a general ROSCA run by shop-keepers like Mokbal Miah, both in Tikkapara, Mohammadpur.

Rickshaw ROSCAs are designed to exploit the daily earning potential of rickshaw drivers, to produce a 'prize' big enough to purchase a rickshaw, to reflect the fact that owner-drivers can save more since they don't have to hire their machine, and to keep 'prized' members from running back to their villages with their prize. They achieve this elegantly, as follows. Rickshaw ROSCAs have a 'manager', often someone in the rickshaw business, such as rickshaw maker, or someone who owns the 'garage'—the outdoor lot where the rickshaws are parked at night. The manager collects a daily savings from each member as the member parks his rickshaw at the end of his day's work. These deposits—often twenty taka per member—are stored by the manager for an average of ten days.

---

[10] Sometimes members who want to invest and gain more than the basic deposit and prize will enrol their children; or else they might simply buy two or more 'names' for themselves. This is common in both ROSCAs and 'fund' *samities*, in Bangladesh and elsewhere.

[11] In India, commercial ROSCAs ('chits') that serve the middle and wealthy classes are seen largely as investment funds, and compete with the stock exchanges.

Each tenth day the draw is held, and if there are twenty-five members the 'prize' will amount to 5,000 taka (twenty-five members *times* ten days *times* twenty taka), enough to buy a good second-hand rickshaw. At this rate it will take twenty-five draws, or 250 days (a little over eight months) to complete the ROSCA.

But rickshaw drivers are often fresh immigrants from the villages. What is to stop them walking away after they win their 'prize'? Apart from its members' propensity to honesty—a strong force—the typical rickshaw ROSCA has two devices to guard against this problem. Members are often obliged to buy a rickshaw with their prize (and sometimes they are obliged to buy it from their 'manager') and this rickshaw can be 'owned' by the general membership until the winner has fulfilled all his obligations. A second clever device is that as soon as a member has won his prize he is asked to increase his savings rate—say from twenty to thirty taka a day. This is seen as fair, since now that he has a rickshaw of his own he pays no hire charges, and can afford to pay more than unprized members. The 'extra' taka is used in one of two ways. In some schemes it is held[12] until the ROSCA is successfully concluded and is then given back to the members concerned, or (sometimes) distributed to all members equally, thereby compensating late winners for their wait, since they contribute least to this fund. In other schemes the extra cash is used to accelerate the ROSCA, since the draw can be held more and more frequently. This reduces the length of time during which prized members have the opportunity to abscond. This innovation of a post-prize premium I have not previously found in the substantial literature on ROSCAs world wide. It may be a contribution from Bangladesh to the development of the device.

ROSCAs can function as a collective, owned and run by a set of equals all well known to each other, as in the case of Mrs Khadeza and her thirteen friends. But more commonly ROSCAs have a manager. In Bangladesh, the transition to manager-run ROSCAs is well under way. We saw this in the case of Nurul Haque's rickshaw ROSCA. Another example is Mokbal Miah. He is a small shopkeeper—his shop is just a tin-sided box standing on public land near a housing complex in Mohammadpur. He has been involved in both 'fund' and ROSCA *samities* for many years, and has gradually

---

[12] In one variant, this fund can be lent out to members on interest.

emerged as a skilful ROSCA manager. Having a manager allows a ROSCA to grow a little bigger—Mokbal's ROSCAs have forty or fifty members—and the members need not have known each other before the ROSCA started. The trust in each other that Mrs Khadeza's ROSCA requires is replaced by general trust in Mokbal Manager's skills. His qualifications are that he is well known, for his little shop has been there for more than eight years, and though he has only a class two pass he is capable of keeping simple books, which we were able to inspect. More than that, he has a strong persuasive personality, excellent for dealing with members who may pay late or try to get round the strict rules. Above all he is pragmatic, and this comes out when we look carefully at how he manages his ROSCAs.

To begin with, he himself chooses the members. He favours people who were in a previous successful ROSCA, or their family members, or new people strongly recommended by previous members. This policy results in a mixed membership of men, women and children, clustered in families. Anyone known or suspected of being unreliable gets politely excluded. Then he has a system of fines for late payers: the size of the fine depends on how late you pay, and if you delay for too long you must leave the ROSCA, in which case you're allowed to take back whatever you have deposited so far, but with no benefit. And anyone who hasn't paid up in full at the time the draw is held is automatically excluded from the lottery that time round. To manage the fact that sometimes members *do* pay late or may even abscond with their winnings, Mokbal needs some reserve cash. To arrange this, he takes a few extra people into membership. For example, if the ROSCA nominally has 40 members each paying ten taka a day with a fifteen-day draw resulting in a prize of 6,000 taka, Mokbal actually recruits around 45 members, but keeps the prize at 6,000 taka. The extra money is used to make up any shortfall in deposits, so that he can guarantee a prize of 6,000 every ten days and keep his word and the confidence of his members. As the extra money builds up, it is held in reserve for pay-outs to drop-outs or against any loss resulting from an absconding winner. If the ROSCA ends successfully, then the extra cash is paid as normal prizes to the 'extra' members. Purists may avert their eyes from these deviations from the elegance of the ROSCA arithmetic, but round the world pragmatism like Mokbal's has allowed ROSCAs to prosper in real life.

Among the ROSCAs that we looked at with special care there are two failures. One of these is a ROSCA that took place in a garments factory in Shamoli. In 1994 a small ROSCA was set up and was successfully concluded, using small deposits paid monthly from wages. A second one followed and was also successful. Then they became ambitious, and muddled the rules. They set out with a much higher monthly deposit, and intended to run it as a 'fund' *samity*, but got cold feet and converted it back to a ROSCA before any loans were made. This led to confusion and loss of confidence. The leader, a floor supervisor, had to bring pressure on members who became increasingly reluctant to pay. Finally the ROSCA collapsed: the fund was returned to depositors as fairly as possible.

Lack of clarity, lack of firm leadership, and muddled rules all lead to losses of confidence and ultimately failure, it seems. But there are other reasons for failure, and one may be scale. We have already seen Nurul Haque's success with his rickshaw ROSCAs. Nearby there is another rickshaw ROSCA manager, Miah Chand, with a string of successful ROSCAs to his credit. But on one occasion Chand Miah went too far. In July 1994 he set up a ROSCA with no fewer than 122 members. Each saved ten taka a day and there was a five-day draw[13], producing 6,100 taka, of which Chand Miah kept 100 taka as his management fee and the prize-winner took 6,000. Prizes had to be spent on rickshaws, and the first thirty-five prize-winners took their 6,000 taka and bought their machines. But with such a large number of members it was impossible for Chand Miah to keep tabs on them all. A few went back home, mainly to Comilla. Chand Miah tried following them to get them to pay their dues, but this proved expensive and largely unsuccessful. When the rest of the members saw that the ROSCA could collapse they precipitated that collapse by withholding their own dues. Chand Miah was left owed many thousands of taka by prized members and in debt by an even larger sum to unprized members. To preserve his reputation and to go on running his other ROSCAs he is having to discharge his debt, and he has sold some rickshaws and a piece of land to cover part of it.

---

[13] There are ways of handling large memberships which don't require draws to be held as frequently as every five days. In Mokbal's ROSCA (discussed above), there are two 3,000 prize-winners who share the 6,000 available each fifteen-day draw. This brings the time taken to run the cycle down from almost two years to a manageable eleven months.

The 'Fund' *Samities*

There are mixed views among managers and members about the comparative merits of ROSCAs and 'funds'. 'Funds' are clearly more flexible, but because they derive this flexibility from the fact that their proceedings are not so 'automatic' and transparent as those of ROSCAs they run a higher risk of muddle or of fraud. My four assistants came unanimously to the conclusion that ROSCAs are 'safer', and that on the whole they run better than 'funds'. We found more failures among 'funds' than among ROSCAs, but this may be misleading, since failed ROSCAs tend to disappear from view immediately, while failed 'funds' can linger on. 'Funds' maintain their popularity and some ROSCA managers are tempted to move into 'fund' management.

Rabia is a part-time 'fund' manager. An ex-NGO worker, Rabia was asked to organise a *samity* by young women whom she had previously had as students in an NGO-run sewing class. They saw her as mature and trustworthy. She agreed, and has now run several. They are fairly relaxed affairs. Rabia keeps numbers small, and personally goes door-to-door to collect the tiny daily savings of two or five taka (she is running two *samities* at present). Many of the women use the *samity* to save up for their children's clothes or schooling, and the names of many of the accounts are those of children or teenagers. Rabia stores the cash in her own home and doesn't use a bank. Members and outsiders can take loans, but Rabia makes all the loan decisions herself. In the troubles of March 1996 many of her members were unable to keep up their savings and some borrowers couldn't make the interest payments on their loans. Rabia weathered the storm. In the end about two fifths of her membership left—the borrowers repaid their loans and the savers took back their savings without any profit—and her two-taka a day *samity* now has just fifteen members left, but is running well, with savings and repayments coming in on time again. When we interviewed these remaining members we found that many had only a rudimentary knowledge of the rules of the *samity*. They didn't all know each other, Rabia being the only link common to all of them. Most knew that there are supposed to be regular meetings of members, but few could say how often, and some had never been to a meeting. But all knew that the *samity* had a fixed life of one year and knew that they could expect to get their savings back then plus a dividend: many were able to guess at the size of the dividend, having been in Rabia's *samities* before. When we asked them what benefit Rabia got from managing the *samity* there was some embarrassment. Many said that Rabia wouldn't expect anything but she'd probably get 'something from each of us' at the end. In fact it is well established that managers get a fixed sum, often fifty or 100 taka, from each member at the successful closing of a 'fund' *samity*. But all sides preserve the fiction that it is unseemly to acknowledge that the manager has any pecuniary interest in the *samity*. This nicety appears to be common in Bangladesh.

The strengths of Rabia's 'fund'—inexpensive informality and flexibility, and most decisions taken by one person of good will—are the weaknesses of many other 'funds'. DH was the 'section-in-charge'—a junior management position—in a garments factory in western Dhaka[14]. Having seen 'fund' *samities* working well in other factories, DH suggested that his co-workers form one, and recommended himself as the cashier. He set a monthly subscription of 100 taka per member. Members were told that the cash would be banked and members would be allowed to take loans of up to 2,000 taka at an interest rate of 5% a month. The *samity* was to run for a year when the accounts would be closed and savings returned and dividends paid. Eighty members joined. For seven or eight months all went well: DH was seen entering the accounts neatly in a cash-book each month, and about twenty-five members got loans varying in size from 500 to 2,000 taka. Then one day DH was transferred to another factory. He handed the accounts over to H, a fellow manager, who found that DH had not been banking the deposits but had used 20,000 taka in his own business. H told the factory Management, who did nothing, and H failed to call a meeting to discuss the matter. The members 'took it very seriously' as one of them described it to us, but what could they do? Well, they stopped making deposits. And there the matter rests. Most members would just like to get their money back, but there seems to be no mechanism to allow this to happen.

Many Bangladeshis—including many of you in the NGO movement—are very sceptical about informal *samities*, arguing that they can't work in a country with such low levels of literacy, such oppressive patron-client relationships, such gullibility and such poverty.

---

[14] Names withheld since the description is somewhat critical.

The story of DH confirms your fears. But the problems can be got round. Look for example at many other garments factory *samities* which *are* running well. Let us get a better idea of the problems by taking a closer look at big *samity* that is about half-way through its annual term, and appears to have a reasonable chance of turning out well.

Asiruddin is an energetic partly-schooled young man who owns some rickshaws, runs a rickshaw 'garage' and trades in eggs. His garage is in a prominent position on the main road into West Agargaon and he is there most of the time. His rudimentary *samity* account book is stored there. The 'fund' he is running at present has about forty members, all men. Many of them are rickshaw drivers who use his garage. Others are local small-time businessmen who run tiny stalls or 'ferry' goods around the slums. The members can often be found lounging in Asiruddin's garage—drinking tea or gossiping. Asiruddin has a track record: he runs ROSCAs and recently completed a successful 'fund' with a six month life. This time he has increased the life of the *samity* to one year but kept the same basic rules. Members save an unvaried 50 taka a week, on a Friday night, and can take small loans on which they pay 2% a week (an APR[15] of around 104%). Savings are made, loans disbursed and repayments and interest payments made *only* at the Friday night meeting, under the eyes of those members who turn up. When we first met Asiruddin in February 1996 the *samity* had just started. A month later seven members had taken loans. In early April Asiruddin showed us his account book. It was neatly kept and showed that most members were saving regularly: for example on the previous Friday 35 out of 40 had saved and Asiruddin expected the remainder to come in within a day or two. Twelve loans were recorded, ranging from 200 to 3,000 taka, and their stated uses were shown—setting up a tea-stall, buying a rickshaw van, stock for a flower-seller and a grocer, and so on.

We began to talk to members. The richest one we found has his own tea-stall, owns more than one rickshaw, and has a minor government job—he says he earns 5,000 taka a month net. The poorest were rickshaw drivers with no other occupation. In the bazaar down the road from Asiruddin's garage we found Alam, who runs a small grocery stall. He was recruited into the *samity* by Asiruddin who, he says, is reliable. There are other financial links between them, since

---

[15] Annual Percentage Rate: the standard annualised way of reporting interest rates.

Asiruddin once hired rickshaws from Alam and they had both been members in another 'fund' *samity* and both are current members of a Proshika *samity*. Alam doesn't know many of the other members of Asiruddin's 'fund'. But he knows the rules and is saving regularly. He has taken a 1,500 taka loan from the *samity* which he has on-lent to a cousin who had taken credit (*'baki'*) from Alam's shop. Alam intends to take no more loans, his main motivation in joining the *samity* having been to save and benefit from the good dividend he expects at the year end. This lump sum will go into his shop. But Alam is himself running a 'fund': a small one with only twenty members and with rather cautious rules—members can take loans of only 70% of their savings deposits to date. Alam clearly treats 'funds' and ROSCAs as part of the normal financial landscape. He is been in many and knows of many more running in the neighbourhood. These days, he says, 'most of them run well', though their rapid growth means that there are still some inexperienced, badly run or fraudulent ones.

Our last look at Asiruddin's *samity* was in late June 1996. Four members had dropped out but had taken their savings before leaving. Some Friday-night meetings hadn't been held, because, we were told, of 'rainy weather', though savings and loan interest payments were still coming in more or less on time. But there'd been problems persuading members to take loans, and though the whole of the fund (now more than 36,000 taka) was out on loan they had had to drop the interest rate on loans, from the very high 2% a week, first to 3% a month, and eventually to 2% a month, to achieve this. The accounts book was still well-maintained as far as savings is concerned, but on the loan side it wasn't fully up to date. Asiruddin didn't appear concerned about this. We give Asiruddin's *samity* a 80% chance of finishing successfully, and look forward to following events as the months go by.

The Wider Perspective

I have studied informal financial services in other South Asian cities [Rutherford 1995 and 1996], which gives me the opportunity to set Dhaka's *samities* in a comparative frame, and to speculate on their future. It looks rosy.

I believe there may be a common pattern as informal financial services develop in an urban area. As ROSCAs and 'funds' arrive and spread by word of mouth around the neighbourhoods there is a high rate of failure and much diversity, as people experiment for themselves with the many variables possible in the running of these *samities*. In this period practices that suit the local conditions—such as the post-prize premium paid by rickshaw ROSCA winners—are discovered and then discarded or perfected. Then two trends begin to emerge. The first is towards professional management, and the second towards neighbourhood-wide and eventually city-wide standardisation. As I see it, Dhaka stands, in 1996, at the start of these trends. The organisation of ROSCAs is increasingly becoming the job of experienced people like Mokbal Miah, and mechanisms for rewarding managers like him are becoming uniform. The same trend towards professional management is occurring in the 'funds', as savers tend increasingly to join 'funds' run by well-known leaders with established track records. Many of these managers, such as Nurul Haque, Miah Chand and Asiruddin, see the management of ROSCAs and 'funds' as a profession in its own right, and a growing share of their own income is derived from this work. Meanwhile, 'funds' may be moving towards uniformity: 'banded' savings contributions (in which everyone saves the same, but there may be provision for a member to hold two or more 'names' if they want to save more), and the twelve-month life are the most evident examples so far.

We may anticipate further moves towards uniformity. In southern India the professional management of ROSCAs reached extreme levels of sophistication many years ago. Called 'chit funds', managed ROSCAs are big business in most towns there, and it is common to see hoardings advertising the services of chit fund management companies. Their activities are regulated by an Act of Parliament[16]. Many banks and other finance houses, including most Co-operative Banks, run chits. Chits have emerged as the favoured form of finance for certain activities—for example buying land is often done through a chit. Chits serve people from all walks of life, including the poor, and are especially popular among the lower middle classes.

In some cities 'fund' rules have become uniform city-wide. In Cochin, for example, in Kerala State, 'annual savings clubs' are known in almost every neighbourhood and take the same form. For every ten rupees saved per week for a year, the saver will get back

---

[16] Chit Funds Act, 1982.

six hundred rupees at the year's end. This represents an annual rate of interest on the savings of 30%, far better than the rate available in any bank. The dividend comes, of course, from interest earned on loans taken from the fund, and these too bear a fixed standardised rate—4% a month. Such clubs offer savers an excellent reward and a very convenient way to save. Many members join to save rather than to borrow. These clubs are run by individuals, neighbourhood groups, churches and temples, or by NGOs, and have managers who receive a standard reward for their services.

Managing the Capacity to Save

What does our brief survey suggest about the market for financial services among the poor and lower-middle neighbourhoods of Dhaka? It suggests that there is a huge market for the safe collection of savings made out of normal income—daily income for many informally employed urban dwellers, monthly for those with waged employment. The most common answer to the question 'why did you join this *samity*?' was 'to save, because it is almost impossible to save at home'. Follow-up questions showed that the intermediation of those savings into a usefully large lump sum is also important. But it is not true that most members want to take that lump sum as a *loan*. This is seen to some extent in the popularity of the lottery ROSCA, where members know from the very start that they must save for a set period, and that since their prize may come at any time they must assume that it may not come until at or near the end of the cycle of draws. In the 'fund' *samities* many respondents appeared happy to wait for the lump sum to arrive after a year of saving, and regarded the availability of loans as an insurance against emergency or otherwise unforeseen expenditure they may be obliged to make, rather than as the main reason for joining. Remember that Asiruddin has had difficulty placing his fund as loans, and has had to reduce interest rates to persuade enough members to borrow enough.

There is little doubt in my mind that our respondents understand these *samities* as being, primarily, about savings rather than about loans. They are right to think so. Similarly, they regard the lump sums of capital that their savings yield as capable of being put to many uses. Questions to our respondents about what they would do with their ROSCA prizes or the yields from their 'funds' produced a wide range of answers. Some will bank the cash as part of a long-term plan to build up a large deposit. Some will convert the cash into other forms of saving, most often gold. Others will spend the money in ways that directly improve their quality of life, such as roof-sheets or furniture or TV sets. Others want to invest in education, training, bribes for getting a job, or the marriages of their daughters. And of course some of the money goes into land purchase or business investments. Among the business investments one use stands out—buying rickshaws or rickshaw-vans—though stock for shops and small trading businesses is also common.

Informal Samities and the NGOs

What can NGOs learn from these *samities*?

Bangladesh's famous financial services NGOs are *lenders*. They lend money to poor and lower-middle class households whom they see as having been hitherto deprived of the chance to borrow at 'reasonable' rates of interest. The literature put out by these NGOs shows that they assume that borrowers will invest in businesses that will produce a steady stream of income out of which the loan will be quickly repaid—one year in most cases. The proportion of borrowers that do invest in such income-generating ways is not known: estimates vary and are hotly contested. But it is clear that many spend their loans in a wide variety of ways that do not produce a stream of income. But as it happens, many NGOs require their loans to be paid back in a series of weekly instalments. This enables borrowers to repay out of normal weekly income, whatever they spend the loan on. Thus for many borrowers NGO group membership is simply another way of exploiting their capacity to save, and is comparable to membership of the *samities*. A series of small regular payments is turned into a usefully large lump sum. For this reason I have described NGO loans as 'advances against savings' [Rutherford 1996 (2)].

By comparison with ROSCAs and 'fund' *samities*, NGO groups reveal some advantages and some disadvantages. On the one hand NGO group membership, being rather more formal, ought to be safer than a *samity*. However, not all NGOs are reliable, and even generally reliable ones may be slow to respond to need—we heard many complaints that one large NGO consistently failed to provide loans at the time it had originally promised. Since NGOs have their own large loan funds NGO loan sizes should be bigger than those available at most *samities*: this is true, but of course loan sizes may be effectively limited by the sum of money the borrower can reliably save out of a normal week's income. On the other hand, arithmetic favours the 'fund' *samity* over the NGO group. This is because in the NGO system a member can turn her savings into usefully large lumps of capital only by going into debt. If I can save 50 taka week I may well prefer to save it in an annual 'fund' and get an unencumbered 3,000 taka back after one year, than spend my fifty taka a week on repaying and servicing for one year an NGO loan of less than 2,300 taka[17]. Only if I am desperate to get my hands on a lump of capital right now (as opposed to waiting for a year) would it make sense for me to choose a loan from an NGO. NGOs seem to assume that virtually all poor people want an immediate loan and are ready to invest it in an income-generating business, but common sense and a little bit of research shows that most would prefer to settle for a safe way of building up their savings, to satisfy a wide range of needs besides business needs. Moreover, many poor people are adverse to debt.

Financial services are essentially about creating lump sums ('liquidity') out of normal income flows [Copestake]. Savings create liquidity by the saver forgoing some income now in return for a lump sum later on, whereas loans create a lump sum now in return for forgoing some of the borrower's future income. A good financial services programme will enable customers to exploit *both* these ways of creating liquidity. At present, NGO schemes offer only the loans option, whereas 'funds' and (to some extent) ROSCAs, offer both options.

'Funds' also offer the opportunity to *store* lump sums. As a 'fund' member I can go on saving as long as I like and store the results as an interest-earning deposit in my *samity*. But in an NGO using Grameen Bank technology I can't store my lump sum. Rather, I have to take it at the beginning of each annual cycle, and spend or invest it there and then. As a result, my investment is often less than optimal.

Thus, for all their undoubted success with their loans, NGO schemes appear clumsy by comparison with 'funds' and ROSCAs when looked at as ways to help poor people manage their savings.

---

[17] An NGO loan of 2,261 taka at a typical 'flat' interest rate of 15% paid back over 52 weeks (principle plus interest) would require a weekly payment of 50 taka (50 taka x 52 weeks = 2,600 taka, divided by 1.15 = 2,261 taka).

One-dimensional, complex, loaded with paperwork, burdened by expensive overheads, slow-moving, inflexible and inherently conservative, NGO schemes, I believe, *need* competition from informal *samities* if they are to improve and prosper.

In time, NGOs may find ways of using their undoubted comparative advantages (good organisation, intelligence, cheap capital, good will, public support, well-intentioned staff, and so on) to fight back to keep or even extend their share of the growing market in financial services for the poor. One way to do this is for them to re-market their services as 'money managers' to the poor. Poor clients would contract to pay in a set amount each week and in return would be allowed to store the deposit and earn interest, *and/or* withdraw it, *and/or* take a loan.

The NGOs would then compete directly with the 'funds'. Moreover, we have seen that when 'fund' *samities* and ROSCAs fail, the reason is often poor or fraudulent book-keeping and weak organisation. Book-keeping and organisation are management skills that NGOs have (or should have) in plenty. By advertising with a slogan like 'all you expect from your 'fund' *samity* plus total security' the NGOs would be well-placed to capture most of the market.

CONCLUSION

Iffath Sharif and Geoffrey D. Wood

Context

IDPAA, Proshika, together with the Credit and Development Forum (CDF) and the Bangladesh Institute of Bank Management (BIBM) is convening two workshops in order to question an increasing orthodoxy in poverty-focused credit. The first one, at which papers included in this volume were presented, focused upon the borrower, and the second will focus upon the lenders. Recent events such as the micro-credit workshops in Reading, UK and in Dhaka in March 1995, sponsored by The World Bank and ODA, UK seemed to be pushing for a stripping away of NGOs' multi-dimensional strategies in favour of a more streamlined focus upon micro-credit delivery. This 'new wave' agenda, in effect, seeks to transform NGOs into Micro-Credit Institutions (MCIs) with two specific objectives of achieving financial sustainability and increased outreach. Such a policy withdraws from a broader political economy analysis of poverty into a narrower, neo-liberal, conception based on poor people's financial liquidity, focusing attention more on poverty alleviation than removal. These concerns are reinforced by the preparatory discussions for the upcoming Micro-Credit Summit in Washington, USA in February 1997, which continues this 'new wave' agenda of restricting the function and purpose of any group formation in poverty-focused development to the policing of micro-credit loans in order to reduce transactions and information costs of lenders.

This 'new wave' agenda is exemplified by the so-called **credit alone** model which is not the only serious model to arise from two decades of experimentation in Bangladesh. The counter model requires a shift from a credit focus towards a **credit-plus** approach, which extends beyond skills and business training to a broader concept of micro-financial services, in which savings feature prominently. The debate between credit alone and credit plus prompts the following questions:

- Are poverty-focused NGOs obliged to expand their micro-credit activities as the only route to their own sustainability?
- Can small NGOs resolve their problems of credit delivery by themselves or must they attach themselves to a larger credit provider, thus prompting the case for an NGO bank providing direct financial services to the clients of NGOs across the sector?
- Can micro-credit ever address the needs of the very poor, and does it at best, confine them to subsistence?
- To what extent are NGOs innovating in financial services and what can they learn from financial experimentation by the poor: e.g. ROSCAs and other savings groups?
- What packages of financial services are actually required by the various groups of the poor, especially in the context of their options in the informal financial sector?
- Last but not least, should financial services be connected to social mobilisation efforts or kept separate, especially in the context of gender rights and poverty removal objectives?

All models of poverty-focused intervention need to acknowledge the heterogeneity of the poor both in terms of their assets, capabilities and psychology. Such models also need to 'prevent' poverty as well as seek to remove it by recognising the significance of 'tomorrow's poor'. Thus a 'lowest-common denominator' small loan approach fails to address the diversity of demand for financial services. Innovations in terms of scale, period and management are required to overcome both poor people's seasonal liquidity at one extreme, and to support larger-scale, technology intensive and employment creating investment at the other. Such a continuum reflects the various forms of need and entrepreneurial capabilities among today's and tomorrow's potential poor; and offers the prospect of graduation via differentiated financial products.

Finally, even a broader concept of financial services needs to be placed within the wider agenda for alleviating and removing poverty: viz. the bargaining strength of the poor in wage negotiations, access to common property, and transforming oppressive cultural norms. Thus, financial services may be a necessary but not sufficient condition for both poverty alleviation and removal. The editors of this volume would like to make a case for the third counter model, **credit with social development,** based on the argument that the value of credit and other financial services for poverty removal (rather than poverty relief) needs to be secured by social mobilisation in order to protect the erosion of income and assist the entry of the poor into new trading, product and labour markets.

Current Achievements and Constraints

Twenty years ago the only credit available to Bangladesh's poor came from informal sources: kin; friends; traders; and money lenders. Today that situation has been changed with large numbers of people able to access credit from semi-formal institutions. The early credit experiments of the Grameen Bank, BRAC and Proshika have translated into major credit operations. The 1980s saw the Grameen Bank and its model dominant in terms of financial flows to the poor. Although this remains the case in the 1990s, there has been increased experimentation and innovation. Some NGOs have moved towards financial service provision in a limited scale and increasingly the poor are establishing their own financial services (ROSCAs and ASCAs).

Achievements

By late 1995, the Grameen Bank and NGOs covered around 25% of target group households, with 16568 million taka (US$ 404 million) in loans outstanding. Coverage varies substantially, however, from area to area and between social groups. Areas with poor roads, low levels of economic activity and weak NCB infrastructure have benefited little from micro-credit. The high expectations held of micro-credit institutions (MCIs) has at times led to the false impression that they can solve the problem of poverty. NGOs and Grameen Bank have performed at much higher levels than government credit schemes and their achievements compare very favourably with all other anti-poverty strategies in the country. Results have been so impressive that Bangladesh is now a centre for the global diffusion of micro-credit ideas, although it is still a recipient of ideas about savings. Currently, most of the savings generated by these institutions tend to take the form of "required fees" for receiving credit, only to be recycled into Revolving Loan Funds. Deposit banking has not been experimented with by these institutions, although other countries have made successful advances in this area.

Constraints

While these institutions have managed to extend micro credit services to the poor (an outstanding achievement), all the major NGOs and Grameen Bank admit that they have serious problems in reaching the hard core poor, resulting in limited coverage. The impact of these institutions on poverty are complex and hard to measure. The evidence from these papers suggests that these institutions have made a significant contribution to poverty alleviation but have had a much more limited impact on poverty removal.

The larger NGOs and the Grameen Bank have required high levels of subsidy to establish their programmes, provided by donors. However, with high recovery rates, increased effective lending interest rates and improved management, subsidy dependence is dropping and is at lower levels than many other anti-poverty strategies.

A number of factors constrain the performance and outreach of NGOs and the Grameen Bank. These include:
- the limited accessibility to mobilised savings (with a few exceptions like Proshika and BURO, Tangail);
- an over-emphasis on credit (the micro-credit mono-culture);
- the lack of investment opportunities for poor people and, especially, the assetless;
- the disadvantaged position of women, who bear the additional cost of securing access to markets and information;
- the inability of these institutions to operate in disadvantaged areas;
- the absence of market demand for services provided by poor borrowers;
- restricted multiplier effects in agriculture and industry;
- natural hazards.

These papers reflect a consensus that while micro-finance can help alleviate poverty and contribute to poverty removal in Bangladesh, significant poverty removal is dependent on economic and social changes well beyond the reach of financial intermediation. Important interventions at the micro level such as micro-finance need to be matched with supportive macro economic policies to create an enabling environment for the poverty removal process.

Thinking Differently about Poverty and Finance

Experiences of most NGOs and financial institutions have been dominated or are increasingly being dominated by micro credit services and limited access to mobilised savings. Several of these papers argue that not only do the poor need opportunities to increase their income, but they also have a crucial need for the smooth management of their household cash flows. The lives of the poor are constantly vulnerable to income erosion as a result of contingencies that may be brought about by structural reasons, sickness, death of an earning household member, and other unforeseen events. Such factors are identified as key elements that may cause the pauperization of today's non-poor, thus significantly increasing the size of "tomorrow's poor". Providing micro credit may increase the scope of the poor to enhance income. It is in providing a varied range of financial services, however, that one may be able to attack reverse mobility of the poor within the poverty bands. Innovation should not only be in refining existing micro-credit services, but also in devising appropriate and diversified financial services to prevent poverty.

There is a need to recognise that the poor are a differentiated group with varied capacities to fight their respective poverty situations. Further analysis of micro finance services is required to tailor them according to the needs of the various groups of the poor.

The New 'New Wave'

The new 'new wave' now refer to flexible financial services which enable poor people to generate lump sums of money, and build up their asset base as well as protect existing assets. As a starting point for the proposed comprehensive route to meeting the financial needs of the various groups of the poor, Rutherford distinguished between two types of financial services:
- Those that assist the poor to build lump sums through forgoing the consumption of income. For example, savings and insurance services allow the amassing of cash at the expense of

current consumption. Loans, on the other hand, provide cash now at the expense of consumption later.
- Those that allow assets to be converted and reconverted into and out of lump sums of cash. For example, the various forms of mortgage and pawning assets.

In reacting to the Rutherford and Wright et al papers, many in the workshop agreed that accessible savings constitute a key financial service and should have the same emphasis as credit in policy making. The poor are good savers as repayers of credit. Being able to save individually not only allows the poor to turn to their savings in times of need, but also helps them build their confidence to borrow money. Existing savings mechanisms with limited or no access tend to reduce the average level of savings per person. Instead, for full utilisation of the savings capacity of the poor, there is a need for their open access to savings and other flexible savings services such as fixed deposit schemes. Borrowers' savings in effect function as collateral for the lender. The introduction of savings services can be used to build capital as well as a cost-effective measure to assure the sustainability of the lender.

Some of the papers have also proposed a revision of various aspects and rigidities of existing models, particularly regarding optimal loan sizes and enforceable joint liability, and weekly repayments. There is a strong recognition that increasing loan size may threaten the joint liability method of reducing lender transactions costs unless borrower income rose steadily to pay the weekly instalments. Given the existing limitations of the market that in effect places a ceiling on borrower income from credit-induced activities, there may be a need for additional sources of income for borrowers to make the weekly repayment instalments. Current credit programmes may then run into the risk of facing self-exclusion by the poorest of the poor. To circumvent this potential outreach problem where appropriate, in project based lending, repayment schedules could be made dependent on project cash flows. Thus, greater flexibility in existing services could be brought about to improve the effectiveness of existing models.

Some Concerns with the New 'New Wave'

Further market research is required to find out the actual financial needs of the poor clients themselves before these financial services can be introduced. Indigenous forms of financial innovation need to be carefully studied for a better understanding of the dynamics involved in financial management by the poor. In this regard the numerous examples of ROSCAs in the slums of Dhaka (described by Rutherford) are a useful source of learning for all.

There is on-going small-scale but significant experimentation. While Proshika already follows a repayment system based on project cash flows, other NGOs are now relaxing the rigid weekly repayment schedule to accommodate larger loans (e.g. BRAC); and offering flexibility within savings accounts (e.g. BRAC, Grameen Bank and ASA). BURO, Tangail, on the other had, have always offered fully-fledged savings services, independent of loans, and is now experimenting with fixed deposit schemes. One has to be aware that the success of such innovations may come about only gradually, as they require a substantial change in the existing borrower-lender relations. Further research is also required to assess the optimal level of flexibility in the financial services that can be offered and the relevant costs associated with it.

Increased expansion of the financial resource base is also an important consideration if large scale financial operations are to be established. Some of the papers are proposing that NGOs should access the local capital market and engage in serious financial intermediation for their poor clients, thereby bringing the poor within the mainstream financial system. This, however, will entail a proper regulatory framework for NGO operation within the formal financial sector.

Further, there is a need to recognise that not all the poor can be expected to be entrepreneurs, engaged in their own profit-making enterprises. Some papers reported evidence from Bangladesh and elsewhere that poor people accept this capability constraint and informally transfer credit between themselves within groups to optimise returns on their loans. Coupled with this notion is the awareness that not all projects are able to bring about great levels of profitability due to overcrowded markets. There is a case, therefore, for larger loans and the identification of productivity-enhancing technologies to support large scale employment-generating activity, where group members pool different inputs (entrepreneurial, shareholding and labour) thus all becoming stakeholders. These initiatives may help to open up wider possibilities to incorporate the poor via credit-supported wage employment.

Conclusion: Borrowers' Sustainability

Much of the donor and academic discourse on micro-credit has been focused upon the sustainability of lenders through the reduction of their information and transaction costs, using the model of joint liability. These papers mainly reflect the practitioner concern with the sustainability of the borrower and the need to reduce their transactions and information costs. It is clear from these papers and the workshop discussions that borrowers' sustainability also relies upon a wider range of financial services than micro-credit, which particularly suffers from project bias. While micro credit models target income-generating projects, wider financial services aim at the smooth management of overall household cash flows. In addition to accessible savings, the poor have requirements for: current account overdrafts to level out consumption fluctuations; lumps of capital at various stages of the family life cycle; and insurance against unpredictable demands on income. At the same time, some in the workshop argued that borrower sustainability, supported by social mobilisation, requires: protection against the erosion of income (e.g. excessive dowry payments); the enhancement of incomes via improved wage bargaining; and access to common property.

These analyses in the workshop support the argument for **credit plus** and **credit with social development** as the precondition for securing the value of **credit alone** strategies. Existing evidence, however, does not give us any clear indication on the exact outcomes of these two models on poverty removal. Nevertheless, it is important to make the distinction between the ingredients of the credit plus and the credit with social development models as they have significant implications on lender's sustainability, which is the theme of a further workshop. It is only by securing the sustainability of lenders that one will be able to secure the sustainability of the borrowers. The debate between the credit plus and the credit with social development models prompts the following questions:

- Which one of the two models promise both the sustainability of the borrower and that of the lender?
- Do the providers of credit plus free-ride on the mobilisation efforts of others?
- Can the two models co-exist and complement each other for the greater interest of the institutions and their borrowers?

- Can there be a fourth model, whereby the diversity of financial services and expenditures on social development are simultaneously maintained by prudential management of funds by NGOs?
- What are the necessary rules and regulations that may facilitate the further strengthening of these models to secure the sustainability of their operations in order to secure the sustainability of their borrowers?

The above concerns lead us to a new set of issues that centre around the NGOs and their future operations given the increasing popularity, quite justifiably so, of the need for financial services, especially savings. For example, can these financial services be provided in a professional manner? The increased diversification of their financial operations may require NGOs to access the local capital markets in order to effectively multiply the deposits of their borrowers. However, what kind of savings guarantee can these NGOs provide to their clients? Do the existing NGOs have the capacity or the legal status to act as financially sound brokers? How can the Bangladesh Bank add value by regulating and supervising these NGO activities? And last but not least, can social development funds be raised from the capital market?

References

Abdullah, T. 1993, "BRAC's paralegal programme: an impact study", Dhaka, BRAC

Abed, F.H. and A.M.R. Chowdhury 1989, "The role of NGOs in international health" Reich M and Marui E (eds.): *International Cooperation in Health*, Dover, Massachusetts, Auburn House Publishing Company

Abed, F.H. and A.M.R. Chowdhury (in press), "How BRAC learned to meet rural people's need through local action" In: Norman Uphoff et al (eds.), *Reasons for Hope: Instructive Experiences in Rural Development*, USA, Kumarian Press

Abed, F.H. and A.M.R. Chowdhury (in press), "Social mobilization for poverty alleviation" Ponna Wignaraja (ed.), *Pro-poor Planning and Social Mobilization: the new social contract between the state and the poor in South Asia*, Paris, UNESCO

ADB January 1994, "Alternative Credit Delivery Systems" Unpublished report for *TRDEP*, ADB, Dhaka

Adener, Shirley and Sandra Burman eds. 1995, "Money-Go-Rounds", Berg, Oxford

Aga Khan Foundation/NOVIB 1993, "Going to Scale: The BRAC experience 1972-1992 and beyond", Toronto

Aghion, Beatriz 1994, "On the Design of Credit Agreement with Peer Monitoring", DEP No 55 London School of Economics

Ahmad, R.S. 1983, "Financing the Rural Poor: Obstacles and Realities", Dhaka, UPL

Ahmed, M., C. Chabbott et al. 1993, "Primary Education for All: Learning from the BRAC Experience", Washington DC, Academy for Educational Development

Bangladesh Bank 1978, "Problems and issues of Agricultural Credit and Rural Finance", Deliberations of the International Workshop on Providing Financial Services to the Rural Poor, Oct. 23-25, Dhaka

Bangladesh Bureau of Statistics (BBS 1995), "Report on the household expenditure survey 1991-92", Dhaka

Bear, M & J Sebstad 1994, "Income and Employment in Bangladesh: Forward Ever, Backward Never" A Review of Ford-Bangladesh's Livelihood, Employment and Income Generation Program, 1974-1994. Dhaka, The Ford Foundation

Bell, C. 1990, "Interactions between Institutional and informal Credit Agencies in rural India" *The World Bank Economic Review*, Vol. 4, No 1, pp. 297-327

Bennett, L and C. E. Cuevas 1996, "Sustainable Banking with the Poor", *Journal of International Development*, Vol. 8 No 2

Bennett, L. 1995, "Donor Approaches to Finance against Poverty: Hydrology or Intermediation" Talking Notes for the Conference on *Finance Against Poverty: Challenges and Advances in Banking with the Poor*, University of Reading, UK, March 27-28, 1995

Besley, T. and S. Coate 1991 "Group Lending, Repayment Incentives and Social Collateral", Woodrow Wilson School of Public and International Affairs Working Paper No 152 Princeton University

Bhattacharya, D. 1991, "Role of policies, Regulations and Institutions in self-employment Promotion in Bangladesh", BIDS: Dhaka

Bouman, F. and J. A. Fritz 1995, "Rotating and Accumulating Savings and Credit Associations: A Development Perspective" in *World Development* Vol. 23 No. 3

BRAC 1994, "A Five-year strategy for BRAC", Dhaka

BRAC 1995, "BRAC policy in providing access of group members to their savings", Dhaka

BRAC 1996, "Research and Evaluation Division Annual Report 1995", Dhaka

BRAC 1996a, "Rural Development Programme Phase III Report", Dhaka

Canadian International Development Agency CIDA 1994, "Poverty Reduction: An Issues Paper", January

Centre For Hi-Tech Info Processing 1995, "Compilation of Human Development Indicators for UNDP", Dhaka

Chambers, R. 1983, "Rural Development: Putting the Last First", London: Longmans

Chambers, R. 1995, 'Poverty and livelihoods: whose reality counts?', *IDS Discussion Paper* No 347, Brighton: Institute of Development Studies, University of Sussex

Chambers, R. 1992, "Rural appraisal - rapid, relaxed and participatory", Brighton, *IDS Discussion paper 311*, University of Sussex

Chen, M.A. 1983, "A Quiet Revolution", Cambridge, MA, Schenkman Publishing Company

Chowdhury, A.M.R., M. Mahmud and F.H. Abed 1991, "Impact of credit for the rural poor: the case of BRAC" *Small Enterprise Development*, 2

Chowdhury, A.M.R. and R.A. Cash 1996, "A Simple Solution: Teaching Millions to Treat Diarrhoea at Home", Dhaka, University Press Limited

Christen, R.P. et al. 1995, "Maximizing the Outreach of Microenterprise Finance: The Emerging Lessons of Successful Programs" (Draft) Paper prepared for Conference on *Finance Against Poverty: Challenges and Advances in Banking with the Poor*, University of Reading, UK, Mar 27-28, 1995

Copestake, J. 1995, "Conference Report: Finance Against Poverty" *Development in Practice* 5 (3): 264-5

Copestake, J. 1995, "The Integrated Rural Development Programme Revisited", University of Bath, UK

Copestake, James 1995, "Poverty-Oriented Financial Service Programmes: Room for Improvement?" Savings and Development Vol. XIX no 4

Cuevas, C. 1995, "Enabling Environment and Micro Finance Institutions: Lessons from Latin America" The World Bank. Paper prepared for Conference on *Finance Against Poverty: Challenges and Advances in Banking with the Poor*, University of Reading, UK, Mar 27-28, 1995

Dichter, T. W. 1996, "Questioning the Future of NGOs in Micro-finance", *Journal of International Development*, Vol. 8 No 2

Doyal, L. 1983, "Poverty and disability in the Third World: the crippling effects of underdevelopment", in O. Shirley ed., *A Cry for Health: Poverty and Disability in the Third World*, Frome, Somerset: Third World Group for Disabled People

Doyal, L. and I. Gough 1991, "A Theory of Human Need", London: Macmillan

Dreze, J. and A. Sen 1989, "Hunger and Public Action", Clarendon Press, Oxford

European Commission 1995, "Bangladesh: Strengthening Poverty Monitoring Systems and Institutions", Dhaka

Evans, T. G., M. Rafi, A. M. Adams, and M. Chowdhury 1995, "Barriers to Participation in BRAC", BRAC Research and Evaluation Division and Harvard Center for Population and Development Studies

Evans, T., M. Rafi, A. Adams, M. Chowdhury 1995, "Barriers to participation in BRAC RDP", BRAC, Dhaka

Evans, T.G., M. Rafi, A.M. Adams and A.M.R. Chowdhury 1995, "Barriers to participation in BRAC RDP", BRAC, Dhaka

Fuglesang, A. and D. Chandler 1993, "Participation as Process - Process as Growth", Dhaka: Grameen Trust

Gibbons, S. S. and S. Kasim 1991, "Banking on the Rural Poor", Center for Policy Research, Universiti Sains, Malaysia

Goetz, A. M. and R. S. Gupta 1994, "Who takes the credit? Gender, power and control over loan use in rural credit programmes in Bangladesh", *IDS Working Paper* No 8, Brighton: Institute of Development Studies at the University of Sussex

Goetz, A.M. and R. S. Gupta 1994, "Who takes the credit? Gender, power, and control over loan use in rural credit programmes in Bangladesh", Brighton, Institute of Development Studies, University of Sussex

Greeley, M. 1994, "Measurement of poverty and poverty of measurement", *IDS Bulletin*, **25** 2 50-8, University of Sussex

Hasan, G. M. and N. Shahid 1995, "A Note on Reasons of Dropout from Matlab Village Organizations", BRAC-ICDDR,B joint research project, Research and Evaluation Division, BRAC

Hashemi, S. M., S. R. Schuler, and A. P. Riley 1996, "Rural Credit Programs and Women's Empowerment in Bangladesh", *World Development*, Vol. 24 No 4

Hashemi, S. M. 1995, IGVGD "Vulnerable Groups Development: Report on the Surveys on Women Participating in the VGD Cycle" Dhaka:WFP

Hashemi, Syed M. 1996, "Rural Credit Programs and Women's Empowerment in Bangladesh", World Development, vol. 24, No., 4

Hashemi, S.M., S.R. Schuler and A.P. Riley 1996, "Rural credit programs and women's empowerment in Bangladesh" *World Development* 24:635-653

Hashemi, S.M. 1996, "Vulnerable groups development: Impact evaluation system", Dhaka, Development Research Centre

Hazell, P. 1992, "The Appropriate Role of Agricultural Insurance in Developing Countries" *Journal of International Development*, 4(6) pp 576-581

Hedrick-Wong, Y., B. Kramsjo & A. A. Sabri 1996, "Experiences and Challenges in Credit and Poverty Alleviation Programmes in

Bangladesh: the Case of Proshika" Paper to Poverty and Credit Workshop, Dhaka, August

Hoff, K. & J.E. Stiglitz 1990, "Introduction: Imperfect Information and Rural Credit Markets-Puzzles and Policy Perspectives" *The World Bank Economic Review* Vol. 4, 1, pp. 235-250

Hossain, M. 1988, "Credit for Alleviation of Rural Poverty: The Grameen Bank in Bangladesh". Research Report 55, IFPRI, Washington DC

Hossain, Mahabub 1988, "Credit for the Alleviation of Rural Poverty: The Grameen Bank in Bangladesh", Washington, D C: IFPRI

Hossain, N. and S. Huda 1995, "Problems of women-headed households" BRAC-ICDDR,B Joint Research Project Working Paper No. 9, Dhaka

Hossain, M. 1985, "Credit for Rural Poor The Experience of Grameen Bank in Bangladesh", Dhaka

Hossain, M. 1992, "Socio-Economic Characteristics of the Poor in Rethinking Rural Poverty A Case for Bangladesh" (eds.) H. Z. Rahman and M. Hossain, Mimeo, BIDS, Dhaka

Hossain, M. 1996, "Rural income and poverty trends", in *1987-1994: Dynamics of Rural Poverty in Bangladesh*, (eds.) H. Z. Rahman, M. Hossain and B. Sen BIDS Mimeo

Hulme, David 1991, 'The Malawi Mudzi Fund: daughter of Grameen', *Journal of International Development*, 3 4, 427-432

Hulme, David and Paul Mosley 1996, "Finance Against Poverty", Vols. 1 and 2, London: Routledge

Hulme, David, Richard Montgomery, and Debapriya Bhattacharya 1996, "Mutual Finance and the Poor: a Study of the Federation of Thrift and Credit Co-operatives SANASA in Sri Lanka", in D Hulme and P Mosley (eds.) *Finance Against Poverty*, Volume 2, London: Routledge, 177-245

International Development Support Services IDSS 1994, "Final Report of Study 1: Alternative Credit Delivery Systems", Manila: Asian Development Bank

Jahan, R. 1989, "Women and Development in Bangladesh Challenges and Opportunities", Dhaka: The Ford Foundation

Jain, P. 1996, "Managing Credit for the Rural Poor: Lessons from the Grameen Bank" *World Development* 24 (1):79-89

Jodha, N. S. 1988, "Poverty debate in India: a minority view" *Economic and Political Weekly* special issue November, 2421-2428

Kamal, A. 1996, "Poor and the NGO Process: Adjustments and Complicities", in *1987-1994: Dynamics of Rural Poverty in*

Bangladesh, by (eds.) H. Z. Rahman, M. Hossain, and B. Sen, BIDS

Khan, K. A. and A. M. R. Chowdhury 1995, "Why VO Members Drop Out", Research and Evaluation Division, BRAC

Khandker, Shahidur R. and Osman H. Chowdhury 1995, "Targeted Credit Programs and Rural Poverty in Bangladesh". World Bank, Washington DC

Khandker, S. A. and O. H. Chowdhury 1995, "Targeted Credit Programmes and Rural Poverty in Bangladesh", paper presented at the workshop on "Credit Programmes for the Poor", Dhaka

Korten, D. 1980, "Community organization and rural development: a learning process approach", *Public Administration Review*, 40

Lewis, W. A. 1954, "Economic Development with Unlimited Supplies of Labour", The Manchester School of Economic and Social Studies, Vol. 22, No 2

Lipton, M. 1983, "Poverty, undernutrition and hunger" *World Bank Staff Working Paper* No 597, Washington: World Bank

Little, A. M. 1957, "A Critique of Welfare Economics", OUP

Lovell, C.H. 1992, "Breaking the Cycle of Poverty: The BRAC Strategy", West Hartford, Kumarian Press

Maddala, G. S. 1983, "Limited dependent and qualitative variables in econometrics" Cambridge University Press

Mahajan, V. & R. Balakrishan 1995, "Innovative Rural Financial Institutions: Rural Banking in Bangladesh" Report of a visit by an Indian Study Team

Maloney, Clarence and Sharfuddin Ahmed 1988, "Rural Savings and Credit in Bangladesh", UPL, Dhaka

Mansell-Carstens, C. 1995, "Las Finanzas Populares En Mexico" Mexico City: Centro de Estudios Monetarios Latino Americanos, Editorial Milenio, ITAM

Maslow, A. 1954, "Motivation and Personality", Harper, New York

Matin, I. 1995, "Group Dynamics in Credit Groups: A Question of Context and/or Design?" University of Sussex D Phil Project Outline

Matin, I. 1996, "Renegotiation of Joint Liability: Notes from Madhupur" Paper to Poverty and Credit Workshop, Dhaka, August

McGregor, J.A. 1988, "Credit and the Rural poor: The Changing Policy Environment in Bangladesh" *Public Administration and Development*, Vol. 8, pp. 467-482

McGregor, J.A. 1992, "Village Credit and the Reproduction of Poverty in Contemporary Rural Bangladesh" Paper from the SASE Annual Conference, Irevine, California

McGregor, J.A. 1994, "Poverty-Oriented Financial Service Programmes: Room for Improvement" (draft), University of Bath, UK

Montgomery, Richard 1995, "Disciplining or protecting the poor? Avoiding the social costs of peer pressure in solidarity group micro-credit schemes", *Journal of International Development*, 8, 2, 289-305

Montgomery, Richard, Debapriya Bhattacharya, and David Hulme 1996, "Credit for the poor in Bangladesh", in D Hulme and P Mosley (eds.) *Finance Against Poverty*, Volume 2, London: Routledge, 94-176

Mosley, Paul 1996, "Metamorphosis from NGO to commercial bank: the case of BancoSol in Bolivia" in D. Hulme and P. Mosley (eds.) *Finance Against Poverty*, Volume 2, London: Routledge, 1-31

Mustafa, S., I. Ara, D. Banu, A. Hossain, A. Kabir, M. Mohsin, A. Yusuf, and S. Jahan 1996, "Beacon of Hope: An Impact Assessment Study of BRAC's Rural Development Programme", Research and Evaluation Division, BRAC, Dhaka

Mustafa, S., I. Ara, et al. 1995, "Beacon of hope: an impact assessment of BRAC's Rural Development Program" BRAC, Dhaka

Mustafa, S., I. Ara, et al. 1996, "Impact assessment study of RDP", Dhaka, BRAC

Osmani, S.R. 1989, "Limits to the alleviation of poverty through non-farm credit", *Bangladesh Development Studies*, Vol. 17 No 4, pp 1-18, BIDS, Dhaka

Osmani, S.R. 1990, "Notes on Some Recent Estimates of Rural poverty in Bangladesh" *Bangladesh Development Studies*, Vol. 18 No 4 BIDS, Dhaka

Otero, M. & E. Rhyne 1994, "The New World of Microenterprise Finance: Building Healthy Financial Institutions for the Poor", Kumarian Press: Hartford, USA

Pitt, Mark M., and Shahidur R. Khandker 1994, "Household and Intra-household Impacts of the Grameen Bank and Similar Targeted Credit Programs in Bangladesh". World Bank, Washington DC

Pitt, M. and S. R. Khandker 1995, "Household and intra-household impacts of the Grameen Bank and similar targeted credit programs in Bangladesh", The World Bank, Washington D.C

Quasem, M. A. 1991, "Limits to the alleviation of poverty through non-farm credit: a comment", *Bangladesh Development Studies*, 129-132

Rahman, H.Z. & M. Hossain (eds.) 1992, "Re-thinking rural poverty: a case for Bangladesh", Analysis of Poverty Trends Project, BIDS: Dhaka, draft

Rahman, H.Z. 1994a, "Rural Poverty Update, 1994" BIDS: Dhaka

Rahman, H.Z. 1994b, "Rural Poverty Update, 1992-93" BIDS: Dhaka

Rahman, H.Z. 1994c, "Rural Poverty Update, 1992: Improvement, But..." BIDS: Dhaka

Rahman, H.Z. 1995, "Rethinking Rural Poverty", Dhaka: University Press Limited

Rahman, R. I. 1996, "Microenterprise Development in Bangladesh" Country Report for the Project Review of Microenterprise Development in Selected DMCs

Rahman, R. I. 1991, "An Analysis of Employment and Earnings of Poor Women in Rural Bangladesh" PhD Thesis of the Australian National University, Canberra

Rahman, H. and M. Hossain eds. (1995), *'Rethinking rural poverty. Bangladesh as a case study',* University Press Limited

Ranis, G. and J. Fei 1964, "Development of the Labour Surplus Economy Theory and Policy", Homewood, Irwin

Ravallion, M. and B. Bidani 1994, "How Robust is a Poverty Profile", *World Bank Economic Review*, Vol. 8 No 1, Washington

Ravallion, M. 1992, "Poverty comparisons: a guide to concepts and methods", *Living Standards Measurement Study Working Paper No 88*, Washington, D C: World Bank

Rawls, J. 1971, "A Theory of Justice", Bellknap Press of Harvard, Cambridge, Mass

Remenyi, Joe 1991, "Where Credit is Due: Income-generating Programmes for the Poor in Developing Countries", London: Intermediate Technology Publications

Robinson, Marguerite 1992, "Rural Financial Intermediation: Lessons from Indonesia", 3 vols., Cambridge, Mass, Harvard Institute for International Development

Rogaly, Ben 1996, "Micro-finance evangelism, destitute women and the hard selling of a new anti-poverty formula", *Development in Practice* 6 May, 100-112

Rutherford, S. 1995, "ASA: The biography of an NGO", Association for Social Advancement, Dhaka

Rutherford, S. 1995, "The Savings of the Poor: Improving Financial Services in Bangladesh."

Rutherford, Stuart 1996, "Almirahs full of pass-books: Cochin", unpublished report for ODA Urban Poverty Office Delhi

Rutherford, Stuart 1995, " Self-Help Savings and Loan Groups: Cuttack, Vijayawada and Calcutta", unpublished report for ODA Urban Poverty Office Delhi

Rutherford, Stuart 1996, "Financial Services for the Poor in Bangladesh", The Independent, Dhaka, 21st and 22nd May

Rutherford, Stuart, "A Typology of Financial Services for the Poor" forthcoming

SAARC 1992, "Report of the Independent South Asian Commission on Poverty Alleviation", Kathmandu

SAARC 1992, "Meeting the Challenge" An Overview of the Report of the Independent South Asian Commission on Poverty Alleviation

Schaffer, P. 1996, "Beneath the Poverty Debate: Some Issues", *IDS Bulletin*, Vol. 27 No 1, Institute of Development Studies, University of Sussex

Schmidt, R.H. & C.P. Zeitinger 1995, "Using Credit-granting NGOs: Where are we now, and where do we want to go from here?" Paper prepared for Conference on *Finance Against Poverty: Challenges and Advances in Banking with the Poor*, University of Reading, UK, Mar 27-28, 1995

Schuler, S.R. and S.M. Hashemi, 1994, "Credit programmes, women's empowerment and contraceptive use in rural Bangladesh", *Studies in Family Planning* 25 April, 65-76

Sen, A. 1981, "Poverty and Famines: An Essay on Entitlements and Deprivation" Clarendon Press, Oxford

Sen, A. 1982, "Choice, Welfare and Measurement" Basil Blackwell, Oxford

Sen, A. 1992, "Inequality Reexamined" Harvard university Press: Cambridge

Sen, A. 1995, "Mortality as an indicator of economic success and failure" London School of Economics, The Development Economics Research Programme

Senge, P.M. 1990, "The Fifth Discipline: The Art and Practice of the Learning Organization", New York, Doubleday/Currency

Sharif, I.A. 1996, "Poverty and Finance in Bangladesh: A New Policy Agenda" Paper to Poverty and Credit Workshop, Dhaka, August

Sharif, I.A. 1994, "Beyond Survival. The Role of Credit in Rural Women's Socio-economic Development" Middlebury College

Sharma, M. and M. Zeller 1996, "Determinants of Repayment Performance in Innovative Group-Based Credit Systems for the Poor: The Cases of BRAC, ASA and RDRS in Bangladesh" *FCND Discussion Paper* No XX IFPRI, Washington DC

Sirohy, H. 1996, "Role of Integration in Rural Development Programmes: especially wage and self employment linkages" MSc Dissertation, University of Bath, UK

Stiglitz, J.E. 1990, "Peer Monitoring and Credit Markets" *The World Bank Economic Review* 4, 3 pp 351-66

Stiglitz, J.E. 1994, "The Role of the State in Financial Markets" *Proceedings of the World Bank Annual Conference on Development Economics* 1993

Streeten, P. 1990, "Poverty Concepts and Measurement" *Bangladesh Development Studies*, Vol. 18, No 3, BIDS: Dhaka

Townsend, P. 1993, "The International Analysis of Poverty", London: Harvester-Wheatsheaf

United Nations Development Programme 1994, "Human Development Report", New York: UNDP

van Koppen, Barbara and Simeen Mahmud 1995, "Case studies of two BRAC tubewell groups", Dhaka: BRAC, mimeo

Von Pischke, JD 1992, "Finance at the Frontier", Washington, DC, World Bank

Von Pischke, J.D. 1995, "Managing the Trade-Off between Outreach and Sustainability by Measuring the Financial Performance of Microenterprise Lenders" Paper prepared for Conference on *Finance Against Poverty: Challenges and Advances in Banking with the Poor*, University of Reading, UK, Mar 27-28, 1995

White, S. C. 1991, "Evaluating the impact of NGOs in rural poverty alleviation", *Bangladesh Country Study, ODI Working Paper*, No 50, London, Overseas Development Institute

Wood, G. D. & R. Palmer-Jones 1991, "The Water-Sellers" Kumarian Press, West Hartford and I T Publications, London

Wood, G. D. et al 1994, "Thinking Differently: Programme Options for the Poor in South Bhola Island, Bangladesh" A review of Action Aid's programme, Dhaka, July

Wood, G. D. 1994, "Bangladesh: Whose Ideas, Whose Interests?" IT Publications, London and University Press Ltd., Dhaka

World Bank 1995, project on, "Credit Programs for the Poor: Household and Intrahousehold Impacts and Program Sustainability"

World Bank 1995, project on, "Credit Programs for the Poor: Household and Intrahousehold Impacts and Program Sustainability"

World Bank 1995, "Micro Credit for Poverty Alleviation" *Report of the World Bank Task Force on Poverty Alleviation Micro Finance Program*

Wright, G., M. Hossain and S. Rutherford 1996, "Flexible financial services for the poor (and not just the implementing organization)" paper presented at the 'International Workshop on Poverty and Finance in Bangladesh: Reviewing Two Decades of Experience'

Yunus, Mohammad 1993, "Hunger, poverty and the World Bank", speech to World Bank Conference on Overcoming Global Hunger, Washington, DC, 29 Nov.-1 Dec

Zaman, H. 1995, "NGOs in Bangladesh: The Topical Issues" The Holiday Newspaper (22 February 1995), Dhaka

Zaman, H., Z. Chowdhury and N. Chowdhury 1994, "Current accounts for the rural poor: a study of BRAC's pilot savings scheme" BRAC Research and Evaluation Division report, Dhaka

Index

Agrani Bank, 49, 208, 209, 211, 214

Association for Social Advancement (ASA), 46, 48, 49, 51, 195, 196, 197, 198, 200, 201, 211, 234-37, 241, 242, 244, 250, 256, 269, 274, 309, 313, 320-22, 327-31, 335, 337, 339, 377

Badan Kredit Kecamatan (BKK), 101, 102, 105, 110, 131

BancoSol, 105, 108, 115, 117, 128, 131, 312

BancoSol, Bolivia, 131

Bangladesh Academy for Rural Development (BARD), 181, 206, 209

Bangladesh Institute of Bank Management (BIBM), 27, 371

Bangladesh Institute of Development Studies (BIDS), 62, 68, 91, 132, 134, 252

Bangladesh Krishi Bank (BKB), 203, 205, 210

Bangladesh Rural Advancement Committee (BRAC), 32, 35, 36, 46-48, 50, 51, 76, 91-93, 101-4, 111-17, 119, 121-25, 128, 129, 131, 132, 171-94, 231-41, 243-47, 250, 252-56, 313, 315-17, 320-22, 327-31, 334, 339, 373, 377

Bangladesh Rural Development Board (BRDB), 121, 132, 140, 146, 206, 207, 209, 234-37, 241, 242, 244

Bank Rakyat Indonesia (BRI), 101, 102, 105, 110, 117, 131, 311

Bolivia, 28, 105, 131, 312

Comilla, 28, 131, 181, 194, 207, 209, 321, 361

compulsory savings, 117, 174, 247, 309, 310, 313, 315, 323, 327, 334, 335, 337

consumption credit, 78, 138, 247

coping capacities, 171, 187, 188

credit, 41, 57, 166, 298, 372, 378, groups, 119, 123, 134

credit with social development, 41, 57, 373, 378

dowry, 34, 35, 110, 161, 163, 164, 166, 169, 170, 201, 378

economic graduation, 68, 70, 71, 75, 81

Employment Income Generating (EIG), 156, 157

empowerment, 47, 84, 87, 89, 120-23, 136, 146, 148-54, 161-68, 172, 176, 184-86, 195, 224, 316

enforcement costs, 51, 121

environment, 63, 64, 78, 80, 84, 101, 136, 145, 169, 172, 173, 185, 190, 201, 336, 375

exclusion, 91, 275, social, 293

female borrowers, 34, 119, 120, 220

financial markets, 45, 62, 71, 74, 111, 137, 140

financial services, 127, 369

formal financial sector, 42, 63, 64, 70, 73-76, 78, 79, 81, 124, 377

gender equity, 173, 186, 190

ghetto credit, 55, 293-95

graduation, 68, 70, 73, 74

Grameen Bank (GB), 27-30, 32-35, 41, 46, 49-51, 53, 57, 61, 97, 101-3, 111-14, 116-23, 128-33, 137, 140, 141, 165, 203, 205, 217-24, 234-30, 230, 232, 256, 267, 285, 286, 290, 317-22, 330, 337, 369, 373, 374, 377

group fund, 118, 219, 222, 225, 266, 320

head count ratio, 64, 88, 89

health, 30, 47, 54, 65-67, 69, 85, 89, 92, 94, 99, 100, 119, 126, 151, 171, 173, 174, 177-79, 184, 186, 190-92, 194, 196, 201, 204, 287, 303, 318, 331, 344, 346

heterogeneity of the poor, 50, 53, 126, 128, 372

imperfect enforcement, 46, 52, 54, 138, 145, 261, 262, information, 52, 145, 261, 262

Income Generation for Vulnerable Group Development (IGVGD), 48, 52, 125, 182-84, 191, 192, 247, 255, 256

income poverty, 44, 98-100

individual savings deposits, 71, 318-20, 322, 337

Indonesia, 28, 97, 101, 102, 105, 112, 131, 311

information costs, 36, 294, 371, 378

insurance, 138

Janata Bank, 209

Kenya, 28, 97, 101, 108, 115, 131

landless, 61, 111, 120, 137, 146, 157-62, 194, 205-10, 218, 222, 251, 252, 255, 287, 300

loan sizes, 54, 72, 77, 80, 114, 188, 197, 236, 239, 240, 244, 245, 266, 268, 269, 280, 282-84, 321, 324, 368, 376

Malawi, 97, 106, 108, 110, 114, 120, 131

Marginal and Small Farms Crop Intensification Project (MSFSCIP), 205, 211, 216

market failure, 46, 145

micro credit model, 378

Micro Enterprises (MEs), 272, 273, 275, 276, 278, 279-81

Micro Finance Institutions (MFIs), 47, 48, 51, 63, 313-15, 322, 326, 332, 335

Micro-Credit Institutions (MCIs), 33, 35, 36, 42, 55, 62-64, 69-71, 73-76, 78-81, 371, 373

Nationalized Commercial Banks, 203

new wave, 31, 32, 371, 375

nongovernmental organizations (NGOs), 28, 32-36, 38, 39, 41-43, 45-49, 51, 53, 55, 56, 61, 83, 84, 91, 93-95, 97, 101, 133, 137-41, 164, 170-72, 179-81, 184, 190, 192, 194, 195, 200, 203, 205, 210-12, 222, 238, 239, 244, 250, 252, 254, 256, 271-75, 277, 278, 280-84, 286, 287, 289, 294, 298, 303, 309, 313, 314, 322, 325, 326, 328-34, 336, 338, 347, 348, 351, 352, 355, 356, 362, 363, 367-75, 377, 379

peer monitoring, 52, 251, 262-65, 298

poor, 49, 67, 68, 76, 78, 90, 101, 104-6, 110, 119, 124, 126, 146, 151, 206, 209, 214, 252, 254, 272, 275, 279, 292, 304, 305, 313, 315, 319, 335, 370, core, 44, 51, 102, 107, 111, 124-28, 130, 153, 154, 170, 171, 253, 256, 293, 374, entrepreneurial, 48, 51-53, 292, 296-99, extreme, 42, 67, 68, 72, 75, 78, 273-75, 277, 279, 280, 281, 290, 292, 295, 296, 301, marginal, 232, 247, ultra, 171, 182, 184, 191, 192, 234, 246

poverty, 33, 40, 43, 62, 64-66, 69, 71, 84, 89, 98, 110, 119, 166, 172, 173, 175, 185, 204, 249, 261, 312, 313, 375, alleviation, 29, 31, 41, 47, 49, 54, 131, 133, 140, 141, 146, 163, 164, 171, 172, 175, 177, 185, 192, 194, 203-8, 210, 212, 213, 224, 272, 274, 284, 289, 293, 295, 319, 371, 373, 374, definition, 85, graduates, 48, 184, 192, line, 43, 64, 67-69, 83, 86-89, 99-102, 104-7, 111, 124, 127, 128, 253, 273, 292, measurement, 42, 85, reduction, 29, 42, 63, 65, 68, 69, 72, 75, 76, 79, 81, 83-85, 89, 90, 94-96, 98-100

production relations, 133, 135, 136, 289

Proshika, 27, 32, 36, 37, 46, 48, 51, 56, 121, 137, 146-49, 151, 152, 154, 155, 158, 160, 161, 164, 169, 170, 250, 256, 300, 354, 365, 371, 373, 374, 377

Rapid Rural Appraisal (RRA), 174

recovery rate, 75, 204, 206, 207, 214, 250, 374

repayment of loan, 29, 30, 43, 49, 51, 52, 72, 75, 77, 80, 88, 116, 117, 120, 150,

158, 175, 178, 191, 196, 197, 199, 204, 208, 217-20, 245, 247, 251, 261, 263-67, 269, 274, 278, 282, 283, 285, 286, 289, 290, 294, 295, 312, 314, 317, 318, 331, 356, 376, 377, period, 219, 220, 278, 283, 295

Revolving Loan Funds (RLFs), 36, 339

Rotating Savings and Credit Associations (ROSCAs), 32, 35, 56, 314, 351-62, 364-70, 372, 373, 377

rural credit, 61, 84, 94, 121, 146, 206, 236

rural development, 27, 85

Rural Development Programme (RDP), 91-93, 101-4, 111, 113, 128, 131, 177, 178, 181, 183, 184, 231, 232, 237-39, 244-47, 250

savings, 31, 32, 34, 35, 37, 38, 45, 52, 53, 55, 56, 69, 71, 72, 76-78, 81, 93, 98, 100, 111, 117, 118, 121, 125-29, 137, 138, 148, 149, 151, 152, 154, 155, 157, 158, 161, 162, 164, 165, 173, 174, 177, 182, 190, 191, 196, 199, 200, 204, 205, 208, 209, 219, 222, 223, 225, 232, 233, 245, 247, 250, 252, 256, 264-66, 269, 270, 279, 285, 286, 290, 299, 300, 309-40, 345-49, 351-54, 356-59, 362-69, 372, 374-79

scaling up, 171, 181, 192, 232

self employment, 54, 271, 272

Sri Lanka, 97, 99, 101, 102, 107, 108, 114, 117, 124-27, 129, 312

sustainability, 38, 71, 177, 200, 378

sustainable development, 154, 155, 164, 190

system loss, 141

targeting loan, 50, 90, 92, 127, 208, 231, 232, 236, 244-46, 256, 291, 292

technology, 47, 77, 171, 174, 175, 179, 191, 220, 241, 271, 283, 369, 372

transactions costs, 28, 30, 52, 54, 217, 251, 297, 376

viability, 30, 36, 37, 43, 44, 51, 55, 62, 71, 72, 74, 98, 102, 105, 117, 120, 127, 222, 251, 289

village organizations, 174, 185, 245

vulnerabilities, 44, 66, 68, 85, 88, 89, 98, 99, 110, 111, 114-16, 129, 246, 295, 302, 303